The Proceedings of the

1st Asia-Pacific Water Summit

The Proceedings of the

1st Asia-Pacific Water Summit

Edited by

Secretariat of the 1st Asia-Pacific Water Summit

Water Security: Leadership and Commitment

3 - 4 December 2007

B-con Plaza

Beppu City | Oita Prefecture

Japan

World Scientific

NEW JERSEY • LONDON • SINGAPORE • BEIJING • SHANGHAI • HONG KONG • TAIPEI • CHENNAI

Published by

World Scientific Publishing Co. Pte. Ltd.
5 Toh Tuck Link, Singapore 596224
USA office: 27 Warren Street, Suite 401-402, Hackensack, NJ 07601
UK office: 57 Shelton Street, Covent Garden, London WC2H 9HE

British Library Cataloguing-in-Publication Data
A catalogue record for this book is available from the British Library.

PROCEEDINGS OF THE 1ST ASIA-PACIFIC WATER SUMMIT

ISBN-13 978-981-283-327-3 (pbk)
ISBN-10 981-283-327-7 (pbk)

Printed in Singapore.

CONTENTS

ACKNOWLEDGEMENT

Mr. Yoshiro Mori
Chair of the Steering Committee of the 1st Asia-Pacific Water Summit
President of the Asia-Pacific Water Forum
Former Prime Minister of Japan

The 1st Asia-Pacific Water Summit was held on December 3rd and 4th in Beppu, Japan. It focused on the theme "*Water Security: Leadership and Commitment*". Heads of state and government, and other national leaders from Asia and the Pacific gathered to raise awareness of water issues and to share in a determination to take concrete actions towards the resolution of water issues.

The opening ceremony was graced by the presence of His Imperial Highness the Crown Prince of Japan and His Royal Highness Prince Willem-Alexander of The Netherlands, Chairman of the UN Secretary-General's Advisory Board on Water and Sanitation. H.E. Mr. Emomali Rahmon, President of the Republic of Tajikistan; H.E. Mr. Tommy Remengesau Jr., President of the Republic of Palau; H.E. Mr. Anote Tong, President of the Republic of Kiribati; H.E. Mr. Ludwig Scotty, President of the Republic of Nauru; H.E. Mr. Emanuel Mori, President of the Federated States of Micronesia; H.E. Mr. Mititaiagimene Vivian, Premier of Niue; H.E. Mr. Apisai Ielemia, Prime Minister of Tuvalu; H.E. Mr. Kinzang Dorji, Prime Minister of the Kingdom of Bhutan; H.E. Mr. Dosbol Nur ulu, Vice Prime Minister of the Kyrgyz Republic and H.E. Mr. Yasuo Fukuda, Prime Minister of Japan also attended the opening ceremony. UN Secretary-General Ban Ki-moon, delivered a message.

After two days of enthusiastic discussions among heads of state and government and other multistakeholder leaders, the world's first summit on water ended with great success.

During the summit, I reaffirmed that sanitation issues, especially those related to toilets, posed a particular challenge to human dignity. In order to save as many children as possible from water-related diseases, it is critical to develop and improve innovative sanitation systems, in addition to conventional ones, based on the new idea of sustainable and recyclable resource management of human waste.

Climate change has become a major topic of discussion. Its most serious impacts are revealed to us through water. While damage caused by water-related disasters such as floods will increase, it is predicted that desertification and water shortages will become more serious. We have come to share a common understanding that, in addition to the reduction of CO_2 emissions, we should focus on adaptive measures toward these problems, since we cannot avoid the negative impacts of climate change.

The heads of state and government, and other multistakeholder leaders pursued their lively discussions over the two days of the summit. Their determination was embodied in the *Message from Beppu*. Based on the recognition that access to safe drinking water and basic sanitation is a basic human right and a basis of human security, it sets new goals to be met by 2025. These goals are one step ahead of the Millennium Development Goals of 2015, and are unprecedented anywhere in the world. They are:

> *To reduce by half the number of people who do not have access to safe drinking water by 2015 and reduce the number to zero by 2025;*
>
> *To reduce the number of people who do not have access to basic sanitation by 2015 and reduce the number to zero by 2025, through the adoption of new and innovative sanitation systems that are not as water reliant as current methods.*

The true success of the summit, however, depends on how we take specific actions towards the resolution of water issues. In that sense, further efforts are required of us.

In the *Message from Beppu*, Japan made a commitment to provide necessary assistance to the region's vulnerable small island states, in order to protect lives and livelihoods from the impacts of climate change. In line with this, Mr. Ichiro Kamoshita, Minister of the Environment visited Tuvalu earlier this year by request of Prime Minister Yasuo Fukuda. We have already initiated a program and experts have been sent by the Japan International Cooperation Agency (JICA) to investigate the current situation in Tuvalu.

I sincerely hope that all participants of the 1st APWS, taking home what they gained there, will initiate real actions. As an organizer of the 1st APWS and a global citizen,

I have reaffirmed my commitment to making an all-out effort, together with the people of the international community, to hand over a rich and beautiful planet to our children.

Lastly, I am deeply grateful to all those involved for their efforts and cooperation towards the success of the summit. I would like to express my heartfelt gratitude to the session conveners, the local people of Oita, the organizers of open events held across Japan, the members of the Oita Prefectural Committee for the 1st APWS, the more than 300 volunteers who supported the summit at the site and at the airport, the officials of the local and central governments, the interpreters who helped to overcome language barriers, and the police and security staff who worked hard to ensure our safety. More than anything else, I would like to express my profound gratitude to all the participants from home and abroad.

I look forward to seeing all of you at the next Asia-Pacific Water Summit.

Professor Tommy Koh
Chair of the Governing Council of the Asia-Pacific Water Forum
Ambassador-at-Large, Republic of Singapore

Since its inception in March 2006, the Asia-Pacific Water Forum has actively sought to generate meaningful progress in solving our region's most critical water challenges by uniting a great number of the region's top water-related organizations.

The challenges are complex, but the underlying message is simple: *We have the will and courage to realize our vision.*

This message was unanimously endorsed by the participants of the 1st Asia-Pacific Water Summit as part of the "Message from Beppu", which, along with the "Policy Brief" and the "Chair's Summary", provide a roadmap for achieving significant progress.

But the 1st Asia-Pacific Water Summit is not an end unto itself. Rather, it is a beginning. The Summit is a critical stepping-stone towards improving the life and livelihoods of our regions people, not least the 700 million still without access to safe and affordable water and the over 1.9 billion people who have no access to safe sanitation.

Specifically, the APWF will build upon the outstanding work already underway under its initial three Priority Themes (Water Financing and Capacity Development, Water-related Disaster Management and Water for Development and Ecosystems) and five Key Result Areas, and will expend its scope of activities to cover even more issues in order to generate new and innovative approaches to solving our region's water problems. The APWF will also continue to raise these water issues to the level of the highest governmental authorities, so that water issues could be ranked high on their national agenda.

The 1st Asia-Pacific Water Summit was a truly historic event. It was a first in a series of events which will continue to bring together political leaders in Asia and the

Pacific to make commitments allowing for more attention to water and more attention to finding solutions. The Summit would not have been possible without the tireless work of a number of people from all across our vast and culturally diverse region. It is our sincerest hope that this momentum will continue to grow and lead to even greater success at the 2nd Asia-Pacific Water Summit.

THE 1ST ASIA-PACIFIC WATER SUMMIT

In the face of the region's pressing water problems, the Asia-Pacific Water Forum (APWF) was created in 2006, under the impetus of late Prime Minister of Japan Ryutaro Hashimoto and is currently presided by the former Prime Minister Yoshiro Mori. The APWF's mission is to provide countries and organizations in the region with a common platform and voice to accelerate the process of effective integration of water resource management into the socio-economic development process of Asia and the Pacific.

In calling for the creation of the APWF at the 4th World Water Forum in Mexico in 2006, the region's water ministers sought to establish an effective mechanism to encourage more collaborative efforts on water resources management and to accelerate the process of effective integration of water resources management into the socio-economic development process of the Asian and Pacific region. The Asia-Pacific Water Ministerial Meeting also called upon the APWF to organize Asia-Pacific Water Summits, to be held, once every 2–3 years.

The APWF is an independent, not-for-profit, non-partisan, non-political network. The goal of the APWF is to contribute to sustainable water management in order to achieve the targets of the MDGs in Asia and the Pacific by capitalizing on the region's diversity and rich history of experience in dealing with water as a fundamental part of the human existence. Specifically, the APWF champions efforts aimed at boosting investments, building capacity, and enhancing cooperation in the water sector at the regional level and beyond.

To achieve this goal, the APWF provides countries and organizations in the Asia-Pacific region with a common platform and voice in articulating the region's strategies and promoting its achievements in solving water problems, including meeting the necessary investment requirements. For this purpose, the APWF comprises a well-coordinated network of member organizations that are able and willing to voluntarily commit their resources in order to deliver high quality products and services that meet the priority needs of policy and decision-makers and practitioners in the water sector. A number of key institutions from the region have committed themselves to supporting the APWF as "lead organizations" responsible for identifying strategies and initiating actions to achieve progress under each of the Priority Themes and Key Result Areas (KRAs) (Annex 3).

The 1st Asia-Pacific Water Summit, under the theme *Water Security: Leadership and Commitment*, was held in Beppu City, Oita Prefecture in Japan, 3–4th December 2007. Targeting decision-makers in the region, the Summit brought together top-level leaders from various sectors including industry, public, education, local organizations, and media, from 56 countries. It is deemed the first ever Water Summit to be convened in the world and was attended by more high-level dignitaries than any other water meeting to date.

Exchanging on the central theme of "Water Security: Leadership and Commitment," Heads of Government made keynote speeches, while a series of sessions provided participants with the opportunity to discuss in further depth some of the issues of concern such as, sanitation, climate change, water financing, water-related disaster management, water for development and ecosystems, developing knowledge and lessons, increasing local capacity, monitoring investments and results, and the CEO water mandate, among others.

The 1st APWS, jointly organized by the Steering Committee of the 1st Asia-Pacific Water Summit chaired by former Prime Minister of Japan Mr. Yoshiro Mori and the Asia-Pacific Water Forum, whose Governing Council is chaired by Professor Tommy Koh was the occasion for the region to focus its attention and energy on the pressing water challenges, to exchange ideas and good practices and to commit to taking action. While a series of concrete initiatives were launched, the participants also unanimously endorsed the Message from Beppu, a call to further the water agenda in the region that offers tangible suggestions to leaders and expresses intent to heighten water issues on the agenda of major upcoming political events. Additional outcomes include the Policy Brief 2007 that outlines the key recommendations issued by the APWF's lead organizations, the Chair's Summary, and the launch of a series of concrete initiatives.

The 1st Asia-Pacific Water Summit Logo

Water is the foundation of life and all forms of social, economic and environment development. The importance of water for life is not just limited to drinking, it also relates to agriculture, sanitation, environment and water-related disasters. Additionally, the issue of water is becoming increasingly crucial because of the impact of climate change on the resources. As a response to this situation, the 1st Asia-Pacific Water Summit will target top-level decision makers, such as heads of government to attempt to solve these issues.

The logo represents drops of water flowing together. Also, the form of the logo reminds us of the traditional Japanese family crest and of the Japanese letter「水」that means 'water.'

Mr. Kazufumi Nagai has created the logo, a symbol for the "1st Asia-Pacific Water Summit". Mr. Nagai is not only a successful artist throughout Japan, he also feels concerned and devoted to the meaning and the significance of the Summit.

The 1st Asia-Pacific Water Summit in Number

Number of Delegations: 231 persons from 36 countries and regions

- Heads of country: 10
- Ministerial level participants: 32

Leaders from various fields: 140 persons (70 persons from abroad)

Total number of participants:

371 people from 56 countries and regions

Attendance at the opening ceremony: 880 (300 from abroad)

Sessions: 10

Concrete initiatives: 17

Journalists: 260 (40 from abroad)

Press conferences: 16

Coverage: 342 articles in newspapers (12/1–12/8)

Open event: 67 (visitors: some 5,000)

Volunteers: over 300

Interpreters: 28

Security: Some 1,500 police officers

Programme

Monday, December 3rd		
Time	**Session**	**Venue**
11:00-12:50	Opening Ceremony	Philharmonia Hall (B1F)
	• Opening Remarks Mr. Yoshiro Mori *Chair of the Steering Committee of the 1st Asia-Pacific Water Summit*	
	• Remarks His Imperial Highness the Crown Prince of Japan	
	• Greetings Mr. Yasuo Fukuda, *Prime Minister of Japan*	
	• Report from the Asia-Pacific Water Forum (APWF) Prof. Tommy Koh, *Chair of the APWF Governing Council*	
	• Keynote Speech His Royal Highness Prince Willem-Alexander of The Netherlands *Chairman of United Nations Secretary-General's Advisory Board on Water and Sanitation*	
	• Video Message Mr. Ban Ki-moon, *United Nations Secretary-General*	
	• Memorial Speech His Imperial Highness the Crown Prince of Japan	
13:00-15:00	Lunch	Reception Hall (2F)
15:00-15:10	Election of Chair, Vice Chairs and Rapporteur/ Adoption of Agenda	Int'l Conference Rm (3F)
15:10-16:40	Speeches by the Heads of Government	Int'l Conference Rm (3F)
16:40-17:00	Coffee Break	
17:00-18:30	Climate Change, Glaciers, and Water Resources in the Himalayan Region Conf. Rm (B1F) CEOs Mandate for Action in Water Rehearsal Rm (B2F) Monitoring of Investments and Results in Water: Key Result Area 4 Meeting Rm 2&3 (B1F)	
18:45	Welcome Reception Hosted by Oita Prefectural Committee of the 1st Asia-Pacific Water Summit	Reception Hall (2F)

Tuesday, December 4th

Time	Session		
9:00-10:30	**Regional Launch of the International Year of Sanitation (IYS) 2008** Philharmonia Hall (B1F)	**The Small Islands' Dialogue on Water and Climate** Conf. Rm (B1F)	**Commitment for Ensuring Water Security in the Aral Sea Basin – Cooperation versus Competition** Int'l Conf. Rm (3F)
10:30-10:50		Coffee Break	
10:50-12:20		**Water-related Disaster Management: Priority Theme B** Conf. Rm (B1F)	**Water for Development and Ecosystems: Priority Theme C** Int'l Conf. Rm (3F)
12:20-14:00	Lunch		Reception Hall (2F)
14:00-15:30	**Leadership for Ensuring Water Security in the Asia-Pacific Region: Knowledge, Financing, and Capacity Development: Priority Theme A and Key Result Area 1** Int'l Conf. Rm (3F)	**Increasing Capacity for Local Actions: Key Result Area 2** Conf. Rm (B1F)	
15:30-16:00	Coffee Break		Int'l Conf. Rm (3F)
16:00-17:30	**Closing Ceremony**		
	• **Reports from Sessions** Session Organizers		
	• **Presentation of Message from Beppu & Chair's Summary** Rapporteur of the 1st Asia-Pacific Water Summit		
	• **Invitation to the 5th World Water Forum**		
	• **From Other Regions**		
	• **Greetings** Representative of the Government of Japan		
	• **Closing Remarks** Mr. Yoshiro Mori *Chair of the Steering Committee of the 1st Asia-Pacific Water Summit*		
18:00	**Farewell Cocktail and Dinner Party hosted by Chair of the Steering Committee of the 1st Asia-Pacific Water Summit**		Convention Hall (B1F)& Reception Hall (2F)

DAY-BY-DAY CHRONICLE

December 2nd, 2007

Dialogue between the CSOs/NGOs and Participants of the 1st APWS

Engaging directly with participants of the 1st APWS, the Dialogue focused on the recommendations from the CSOs and NGOs, a Message to be addressed to the participants of the 1st APWS (Annex 5).

December 2nd, 2007

Third Steering Committee Meeting of the 1st APWS

The members of the Steering Committee of the 1st APWS convened and adopted the programme of the 1st APWS.

December 3rd, 2007

Opening Ceremony

All participants were invited to the Opening Ceremony, which included remarks and speeches by Mr. Yoshiro Mori, Chair of the Steering Committee of the 1st APWS, His Imperial Highness the Crown Prince of Japan, Mr. Yasuo Fukuda, Prime Minister of Japan, Professor Tommy Koh, Chair of the Governing Council of the APWF, His Royal Highness Prince Willem-Alexander of The Netherlands, Chairman of the United Nations Secretary-General Advisory Board on Water and Sanitation, and a message from Mr. Ban Ki-moon, United Nations Secretary-General.

December 3rd, 2007

Election of Chair, Vice Chairs and Rapporteurs/Adoption of Agenda

Following the opening ceremony, the participants of the 1st APWS proceeded to adopt the agenda and the elections:

Chair

Mr. Yoshiro Mori, Chair of the Steering Committee of the 1st Asia-Pacific Water Summit

Vice Chairs

H.E. Mr. Emomali Rahmon, President, Republic of Tajikistan

H.E. Mr. Anote Tong, President, Republic of Kiribati

H.E. Lyonpo Dr. Kinzang Dorji, Prime Minister, Kingdom of Bhutan

Hon. Mr. Lim Kean Hor, Minister of Water Resources and Meteorology, Kingdom of Cambodia

H.E. Mr. Hu Siyi, Vice-Minister of Ministry of Water Resources, People's Republic of China

Rapporteurs

Ms. Erna Witoelar, Vice Chair of the Governing Council of the Asia-Pacific Water Forum

Mr. Ravi Narayanan, Vice Chair of the Governing Council of the Asia-Pacific Water Forum

December 3rd, 2007

Speeches by Heads of Government

Speeches on water issues and water-related initiatives in each country were delivered by:

- H.E. Mr. Emomali Rahmon, President of the Republic of Tajikistan
- H.E. Mr. Anote Tong, President of the Republic of Kiribati
- H.E. Lyonpo Dr. Kinzang Dorji, Prime Minister of the Kingdom of Bhutan
- H.E. Mr. Tommy Remengesau Jr, President of the Republic of Palau
- H.E. Mr. Ludwig Scotty, President of the Republic of Nauru
- H.E. Mr. Emanuel Mori, President of the Federated States of Micronesia
- H.E. Mr. Mititaiagimene Vivian, Premier of Niue
- H.E. Mr. Apisai Ielemia, Prime Minister of Tuvalu
- H.E. Mr. Dosbol Nur ulu , Vice Prime Minister of the Kyrgyz Republic
- Concluding remarks of Speeches by the Heads of Government

December 3rd, 2007

Welcome Reception

The Welcome Reception was hosted by the Oita Prefectural Committee of the 1st APWS, at which Mr. Katsusada Hirose, President of the Oita Prefectural Committee of the 1st APWS, welcomed the participants to the Summit in his opening remarks, followed by a Japanese-style toast (*Kagami Biraki*) performed by Mr. Yoshiro Mori, Chair of the 1st APWS. The evening carried on with some *Yoshihirogaku* entertainment.

December 3rd and 4th, 2007

Sessions

Ten sessions were organized by various stakeholders, which gave participants opportunities to engage directly and in further depth on themes of critical importance for water policy in the region.

- The session "Climate Change, Glaciers, and Water Resources in the Himalayan Region" focused on the necessity of regional cooperation for the assessment and monitoring of disasters caused by retreating glaciers and their impact on water resources.

- At the session "CEOs Mandate for Action in Water", business leaders in the region exchanged views and best practices on various efforts relating to water issues.

- At the session "Monitoring of Investments and Results in Water", participants discussed how to integrate monitoring of investments in official policies.

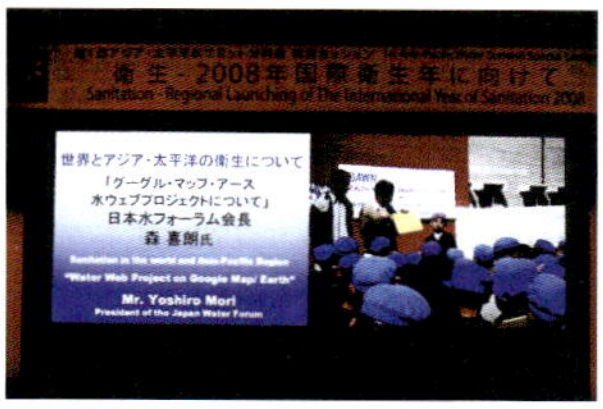

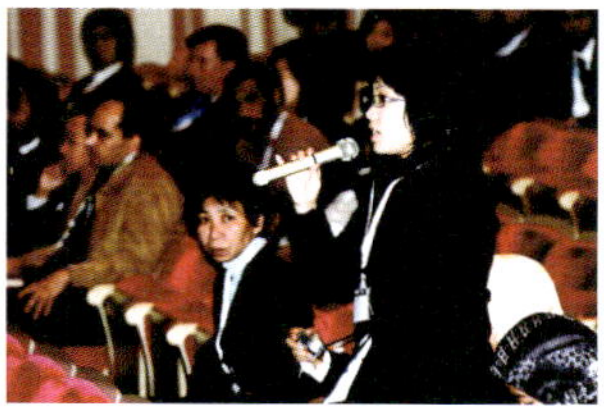

- At the "Regional Launch of the International Year of Sanitation (IYS) 2008", His Royal Highness Prince Willem-Alexander of The Netherlands delivered the inaugural address, followed by discussions on the roles of each sector. The President of the Japan Water Forum, Mr. Yoshiro Mori launched the Water Web Project on Google Earth.

- At the session "Small Islands' Dialogue on Water and Climate", heads of government stressed the need to tackle serious water-related disasters and the difficulty in securing water resources.

- At the session "Commitment for Ensuring Water Security in the Aral Sea Basin – Cooperation versus Competition", H.E. Mr. Emomali Rahmon, President of the Republic of Tajikistan, delivered the keynote speech. Discussions were held on the management of transboundary rivers and the necessity of cooperation between concerned countries.

- The session "Water-related Disaster Management" appealed for participants to recognize the adaptation to the risks of climate change as a top priority issue.

- The session "Water for Development and Ecosystems" focused on how to secure sufficient water for the benefit of both development and ecosystems from a long-term perspective. Participants were made aware of the importance of enhancing governance at the local level.

- At the session "Leadership for Ensuring Water Security in the Asia-Pacific Region: Knowledge, Financing, and Capacity Development", it was recognized that investing in water is investing in poverty reduction. Discussions ensued on various financing mechanisms between governments and water users.

- At the session "Increasing Capacity for Local Actions", a concrete initiative for creating Capacity Building Hubs was introduced.

December 4th, 2007
Closing Ceremony

Following session reports presented by each session organizer, representatives of the World Water Council and the Turkish government delivered speeches on future collaboration initiatives leading up to the 5th World Water Forum. Representatives from the American and the European regions made speeches on interregional cooperation in the future, following which, PUB Singapore announced that Singapore was considering hosting the 2nd Asia-Pacific Water Summit. The rapporteurs proceeded to read out the Chair's Summary and Message from Beppu, which was unanimously endorsed. Mr. Tetsuzo Fuyushiba, the Minister of Land, Infrastructure, Transport and Tourism of Japan, delivered a speech on behalf of the Japanese government. The ceremony concluded with an address by Mr. Yoshiro Mori, Chair of the 1st APWS, wherein he thanked all who had supported the Summit and stated his expectations for future activities.

December 4th, 2007
Farewell Cocktail and Dinner Party

The 1st APWS concluded, all participants were invited by the Chair of the 1st APWS, Mr. Yoshiro Mori. The Farewell Cocktail provided the opportunity to enjoy the 1st APWS exhibition and was followed by a sitting buffet style dinner. The dinner was officially closed with a Japanese Noh Performance.

December 3rd and 4th, 2007

Social Programme and Events

The participants of the 1st APWS and their spouses were provided with a rich programme of social events, including the Tea Ceremony, the Global Tower Tour, the Projection of the Asian Broadcasting Union TV documentary on Water, and an Oita Excursion Programme that included visits to the famed Beppu hotsprings and other sites in and around the city.

December 1st to 5th, 2007

In parallel to the Summit, a series of "Open Events" were organized, to which all stakeholders were invited to facilitate mutual communication and collaboration between all stakeholders in the region for the resolution of water issues. Open Event organizers were invited to present specific recommendations or initiatives to resolve the water issues of the region which were transmitted to the Summit participants.

December 1st to 4th, 2007

Exhibition and Trial of a Bio-Toilet

The Japan Toilet Association and the Japan Water Forum, in cooperation with the Mikasa Corporation, installed sample bio-toilets within the Summit venue for use by participants. This exhibition served the purpose of raising awareness with regard to the importance of sanitation and hygiene among the international community, to improve the lives of the 2.6 billion people who do not have access to basic sanitation in the world.

OPENING CEREMONY

- Opening Remarks by Mr. Yoshiro Mori, Chair of Steering Committee of 1st Asia-Pacific Water Summit/Former Prime Minister of Japan
- Address by His Imperial Highness the Crown Prince of Japan
- Greetings by H.E. Mr. Yasuo Fukuda, Prime Minister of Japan
- Report from the Asia-Pacific Water Forum by Professor Tommy Koh, Chair of the Governing Council of the Asia-Pacific Water Forum
- Keynote Speech by His Royal Highness Prince Willem-Alexander of The Netherlands, Chairman of United Nations Secretary-General's Advisory Board on Water and Sanitation
- Message by H.E. Mr. Ban Ki-moon, United Nations Secretary-General
- Commemorative Lecture by His Imperial Highness the Crown Prince of Japan

Opening Remarks

Mr. Yoshiro Mori
Chair of the Steering Committee of the 1st Asia-Pacific Water Summit
Former Prime Minister of Japan

Good Morning, honoured guests, distinguished delegates, ladies and gentlemen.

It is a great pleasure and honour for me to address you at the opening of the 1st Asia-Pacific Water Summit. I would like to express my profound gratitude to His Imperial Highness the Crown Prince of Japan, and His Royal Highness Prince Willem-Alexander of The Netherlands, Chairman of the UN Secretary-General's Advisory Board on Water and Sanitation, for attending. I would also like to thank President Emomali Rahmon of the Republic of Tajikistan; President Tommy Remengesau Jr. of the Republic of Palau; President Anote Tong of the Republic of Kiribati; President Ludwig Scotty of the Republic of Nauru; President Emanuel Mori of the Federated States of Micronesia; Premier Mititaiagimene Vivian of Niue; Prime Minister Apisai Ielemia of Tuvalu; Prime Minister Kinzang Dorji of the Kingdom of Bhutan; Vice Prime Minister Dosbol Nur ulu of the Kyrgyz Republic. Thank you as well to Prime Minister Yasuo Fukuda, who is here on behalf of Japan, the host country.

I am also grateful to all the people involved, for their efforts and cooperation towards the success of the 1st Asia-Pacific Water Summit. I would like to express my sincere thanks to the conveners of the sessions, organizers of the open events held in Oita and many other places across Japan, members of the Oita Prefectural Committee for the 1st APWS, concerned local governments, NGOs, and members of the public.

The 1st Asia-Pacific Water Summit is a historic event, dedicated exclusively to the resolution of water issues. This summit, focusing on the theme "Water Security: Leadership and Commitment", is expected to provide a good opportunity for governments and other national leaders to raise the awareness of water issues and share a determination to take concrete actions towards the resolution of water issues.

Water problems in the Asia-Pacific region are severe. This region is home to 3.7 billion people, or 60% of the world's population. One out of five (0.7 billion) people in the region have no access to safe drinking water, and half of the population (1.9 billion) has no access to sanitation facilities. We are faced with a critical situation. In addition, we have suffered tremendous losses caused by water-related disasters. Eighty percent of worldwide deaths resulting from water-related disasters have occurred in this region.

In areas where there is no access to safe drinking water, women and children are forced to spend hours collecting water, wasting precious time which could be used for education and social activities. In places where there is no adequate sanitation, people are susceptible to epidemics and other health problems. In places vulnerable to water-related disasters, people tend to lose their desire to work, hampering economic development. With regard to infectious diseases, an area in which I have been involved for a long time, babies who are healthy at birth later contract HIV from breast milk due to the lack of safe drinking water required to prepare baby formula.

In areas of rapidly growing economies and populations, it is necessary to allocate limited water resources appropriately between agriculture, industry, households, environment users, and other sectors. Needless to say, it is critical to secure close coordination of the use of international rivers among riparian countries. In this context, water is a key factor that affects not only local stakeholders but also the survival of the countries themselves. Therefore, national leaders must be in the forefront of water-related initiatives.

In order to focus our discussions during the preparatory meetings for the summit, water issues in this region have been classified into three priority themes: *Water Financing and Capacity Development*; *Water-related Disaster Management*; and *Water for Development and Ecosystems.* The outcomes of the discussions will be presented at sessions devoted to working out concrete initiatives.

In recent years, climate change has sparked great controversy across the world. Its serious impact will be revealed to us through water. While damage caused by water-related disasters such as floods will increase, it is predicted that desertification and water shortages will also become more serious. Intensive discussions will take place on climate change and water in Asia and the Pacific region, especially on glacial melting and the breaking of glacial lakes in the Himalayas, as well as the problem of rising sea levels in the small Pacific island states. Now that we cannot avoid the negative impacts of climate change, our strategy must be to focus on how to adapt to these impacts.

The year 2008 has been declared as the International Year of Sanitation by the United Nations. Among sanitation issues, toilets poses a particular challenge to human dignity. In order to save as many children suffering from water-borne diseases as possible, it will be necessary to discuss seriously the improvement and development of sanitation systems, based on the new idea of human waste as a sustainable and recyclable resource.

The initiative of the Asia-Pacific Water Summit was first presented at the 4th World Water Forum in March, 2006. Shortly after that, former Prime Minister of Japan Ryutaro Hashimoto suddenly passed away. He had dedicated all his energies to the resolution of global water issues and was also the founder of the summit.

I decided to carry on with Mr. Hashimoto's strong determination to contribute to the resolution of global water issues. I have made the greatest possible effort to ensure the success of the summit. I hope my knowledge and long experience in politics including as a prime minister will be of some help in the resolution of global water issues. I believe my mission is to become foster father of Mr. Hashimoto's legacy.

Finally, I would like to express my deepest gratitude to government leaders and all those who have traveled far to attend this summit. I trust these two days will yield valuable results, and I sincerely hope that we will be able to hand over this rich and beautiful planet to our children by overcoming our serious water problems.

Address

His Imperial Highness the Crown Prince of Japan

Your Royal Highness, Excellencies, Distinguished guests, Ladies and gentlemen.

It gives me great pleasure to be with you at this 1st Asia-Pacific Water Summit which is being held here in Beppu City, Oita Prefecture with so many participants from the Asia-Pacific region as well as from other places around the world.

Water is essential for all life on earth. However, this indispensable resource is unevenly distributed throughout the world from a geographical point of view as well as from a time point of view. Thus, water is at the root of all sorts of problems due to exploding populations, rapid economic growth and global warming.

It is for this reason, international efforts at every level, from the humblest village to the United Nations, are essential in addressing this issue. I myself became personally concerned with this question when I participated in the 3rd and 4th World Water Forums, where I listened to the earnest discussions of the participants, and have since witnessed some concrete actions that have evolved from these discussions.

I understand that this Summit came into being during the preparation process of the 4th World Water Forum. As one who has had a strong interest in water, I am especially delighted to be here with you today.

Water issues, however, are not being solved as urgently and effectively as they ought to be. Sanitation especially lags seriously behind other efforts.

The situation in the Asia-Pacific region does not allow us to be optimistic. The region, which is home to about sixty percent of the world's people, possesses only about forty percent of the world's water resources. As of 2004, there were seven hundred million people who had no access to safe drinking water and 1.9 billion who were without basic sanitation. In this respect, our region is in the most serious situation in the world, especially in providing sanitation. Another critical problem is the frequency and magnitude of natural disasters caused by water, and eighty percent of all fatalities in these disasters occur in the Asia-Pacific region.

About two weeks ago, the cyclone that struck Bangladesh took the lives of many people, and caused huge damage to homes, domestic animals and crops. I extend my sincere condolences to the victims, and I share the sense of immense loss and grief of those who survived, and pray that they will be able to rebuild their lives and homes as soon as possible.

Water also has influence on various other problems such as women's empowerment and the achievement of universal primary education, and therefore is a vital issue that affects the very survival of every nation.

It is therefore all the more significant that heads of state and representatives of governments, international organizations, civic groups, the business sector and academia get together to seriously address the common regional problems of water and sanitation, in the search for lasting comprehensive solutions.

I have accepted the request of the United Nations Secretary General Mr. Ban Ki-moon to serve on his Advisory Board on Water and Sanitation as its Honorary President. I am aware of the responsibility that comes with this honor. The Board, under the leadership of its late chairman Mr. Ryutaro Hashimoto, launched the Hashimoto Action Plan and is forging ahead towards achieving the Millennium Development Goals. I will do my best to learn more about the global issues of water and sanitation, and to work with His Royal Highness Prince Willem-Alexander of The Netherlands, the Board's chairman, and its other members to contribute to its activities in my capacity as Honorary President.

In closing, I want to voice my sincere wish that the Water Summit will mark a major step forward toward resolving the global issues of water here in the Asia-Pacific region and throughout the world.

Thank you.

Greetings

H.E. Mr. Yasuo Fukuda
Prime Minister of Japan

Good morning, honoured guests, distinguished delegates, ladies and gentlemen.

At this, the opening of the 1st Asia-Pacific Water Summit, it is a great pleasure and honor for me to deliver a speech on behalf of the Japanese government. First of all, I would like to express my profound gratitude to His Imperial Highness the Crown Prince of Japan and His Royal Highness Prince Willem-Alexander of The Netherlands for attending. I would also like to extend a most hearty welcome to the presidents, prime ministers, representatives of international organizations and all the participants for coming here to Beppu.

About 230 years ago, a Japanese scholar named Shihei Hayashi wrote, "The waters flowing under Nihombashi in Edo and the waters in the rivers of China and Holland are one stream without any barrier." Though Japan is an island nation surrounded on all four sides by water, it is linked to the world through the ocean. When Shihei Hayashi wrote this, he was warning the Japanese people about the risk of threat from other nations. Today, however, Japan shares common interests and lives together with other nations, connected by the water of the ocean. It is very meaningful for us to discuss important water issues here in Beppu, which is proud of its rich, clean water and its natural environment.

Water Issues Asia is Facing

Needless to say, water is an essential resource for the sustenance of all life, the conservation of the environment and ecosystems, and the progress of economic activity. Unfortunately, we are facing serious problems due to water shortages, and water itself can bring various disasters to humanity. Water-related disasters caused by climate change and environmental destruction threaten not only our lives but also our social structures. Today, rapidly developing countries in Asia and the Pacific region are enjoying prosperity, but at the same time they have various water-related problems. Considering the fact that most of the global water problems occur in this area, we have to admit we are facing imminent crises.

Three major challenges in this region may be identified. One is securing safe and hygienic water resources. According to the United Nations World Water Development Report (2003), about 700 million people in the Asia-Pacific region have no access to safe drinking water. It is expected that water demand will greatly increase in this region due to rapid economic development and population growth.

Another challenge is water-related disasters. Water-related disasters are already increasing, presumably due to the impact of climate change. This impact will become greater in the near future. We should take immediate measures to mitigate water-related disasters.

The final challenge is the issue of water for agricultural use. A lot of people across the world are suffering from hunger and malnutrition due to food shortages. More than 500 million of them are concentrated in the Asia-Pacific region. In order to supply sufficient food to these people, it is essential to secure enough water for agricultural use.

Japanese Assistance in the Water Sector

How should we deal with these challenges? Water is a recirculating resource. Therefore, it can be used in a sustainable manner if we utilize it wisely. The integrated management of water resources is very important for effective and efficient use of water. Japan, with accumulated experience and technologies in this field, has been fostering international cooperation. For example, Japan has provided assistance to the Mekong River Commission, which manages water resources of the Mekong River. Planners there are now taking into account both basin development and environmental conservation.

Climate Change

We are faced with the enormous problem of climate change. The Intergovernmental Panel on Climate Change Fourth Assessment Report has identified various effects of climate change: worsening water shortages for several hundred million people; an increasing risk of floods and storm winds; damage to agricultural production; and the risk of losing national territory. In short, the impact of climate change will have far-reaching ramifications.

In order to solve the issues associated with climate change, we must establish an international framework that will lead to the reduction of greenhouse gas emissions across the world. Therefore, in May this year, Japan announced a plan called "Cool Earth 50", which aims to halve the emission of greenhouse gases by 2050. In addition, as a means of promoting the establishment of an international framework, Japan has proposed the establishment of a special financial mechanism. This is aimed at giving assistance to developing countries which are trying to achieve both economic growth and the reduction of greenhouse gas emissions.

G8 Hokkaido Toyako Summit

At the G8 Hokkaido Toyako Summit, which will be chaired by Japan, issues of the environment and climate change will be one of major agenda items, along with issues concerning development and in Africa. I am sure that productive discussions at the 1st Asia-Pacific Water Summit will give strong momentum and provide excellent advice to the G8 Summit.

International Year of Sanitation

Finally, I would like to mention the issue of water and sanitation. On November 21st in New York, the launching ceremony of the "International Year of Sanitation 2008" was held. The regional launch of the IYS will be conducted tomorrow during this Summit. The IYS was one of the recommendations proposed by the UN Secretary-General's Advisory Board on Water and Sanitation. I would like to pay homage to His Royal Highness Prince Willem-Alexander of The Netherlands, who succeeded the former Chairman of the Board Ryutaro Hashimoto, for his remarkable contribution to water and sanitation issues. I am very pleased to know that His Imperial Highness the Crown Prince of Japan has assumed the position of Honorary President of the Board. It is well known to the Japanese people as well as those abroad that the imperial family including the Crown Prince have a keen interest in nature and the environment, and have been involved in its study. I hope that with the cooperation of the Crown Prince of Japan and Prince Willem-Alexander of The Netherlands, the

UN Secretary-General's Advisory Board on Water and Sanitation will make further progress towards the resolution of water and sanitation issues.

Conclusion

Oita Prefecture has a long tradition of using water wisely. The city of Beppu is one of the most famous hot spring resorts in Japan, and has used its hot springs in various ways. It is famous in connection with water. Therefore, I hope the 1st Asia-Pacific Water Summit to be held here will be a good opportunity to cultivate friendships and promote international cooperation for the resolution of water issues.

In conclusion, I would like to say thank you to Mr. Yoshiro Mori, former Prime Minister of Japan, the Chair of the Steering Committee of the 1st APWS, and the President of the Japan Water Forum, and to all the people who have committed themselves to this Summit.

Thank you for your kind attention.

UNOFFICIAL TRANSLATION

Report from the Asia-Pacific Water Forum

Professor Tommy Koh
Chair of the Governing Council of the Asia-Pacific Water Forum
Ambassor-at-Large, Republic of Singapore

Your Imperial Highness, the Crown Prince of Japan, Your Royal Highness, the Crown Prince of Orange of the Netherlands, Chairman, UN Secretary-General's Advisory Board on Water and Sanitation.

Your Excellency, Prime Minister Fukuda, Your Excellency, former Prime Minister Mori, Your Excellencies, the Heads of State and Government, Ministers, Excellencies, Distinguished Guests, Ladies and Gentlemen.

Welcome by Governing Council

On behalf of the Governing Council of the Asia Pacific Water Forum, it gives me great pleasure to extend a warm welcome to all of you, especially to the many friends from outside Japan. It is an honour for me and for my two Vice-Chairs, Ambassador Erna Witoelar of Indonesia and Mr Ravi Narayanan of India, to welcome you all to the first Asia Pacific Water Summit.

This historic event would not have been possible without the tireless work of a number of people from across our vast and diverse region. Such a level of intra-regional cooperation is a remarkable achievement of the Asia Pacific Water Forum and of its ever-growing network of members and friends.

What is the Asia Pacific Water Forum?

But what exactly is the Asia Pacific Water Forum or APWF as it has come to be called?

It is an open, inclusive, non-partisan, non-political, multi stakeholder network. Every government, local and municipal authorities, academic and civil society organization, which has an interest in water, is welcome to join the Asia Pacific Water Forum.

Thank the JWF and the Secretariat

We want to thank former Prime Minister Mori for allowing the Japan Water Forum to serve as the Secretariat of the Asia Pacific Water Forum. I want to thank three particularly outstanding members of the Secretariat: Mr Hideaki Oda, Ms Noriko Yamaguchi and Ms Taeko Yokota. Please join me in giving them a warm round of applause.

Homage to PM Hashimoto

Some of you will recall the Joint Declaration issued by the Water Ministers of the Asia-Pacific Region, on 21 March, 2006, during the Fourth World Water Forum held in Mexico City. The Joint Declaration called for the creation of the Asia Pacific Water Forum and provided in its mandate, to work "*in complete solidarity to identify and adopt solutions to water issues in the region.*" The fact that this Summit is being held less than two years after the adoption of that declaration is a testimony to the hard work and effort put in by all the APWF members. The founding of the APWF was an initiative of the late Prime Minister Hashimoto and I would like to dedicate this Summit to his memory.

Meeting the MDG Target 10

The Asia Pacific Water Forum aims to contribute to sustainable water management and the achievement of the UN's Millennium Development Goals and Targets in the Asia Pacific. Target No. 10 is to reduce by half the number of people in the world without access to safe drinking water and modern sanitation by 2015. ASEAN, at its recent Summit in Singapore, committed itself to "endeavour to reduce by half of the number of people without sustainable access to safe drinking water by 2010". In the Asia Pacific, in spite of our growing prosperity, 700 million people do not have access to safe drinking water and 1.9 billion people do not have access to modern sanitation. The situation is unacceptable. We have the means to make safe water and modern sanitation a basic human right of all the peoples in this region. Our collective ambition is to ensure that water policy, reform, investment and performance are high on the agendas of our region's leaders.

APWF's Two Objectives

The APWF's objectives are: (i) to provide countries and organisations in the Asia-Pacific region with a common platform and voice to articulate the region's strategies and promote its achievements in solving water problems and (ii) to showcase leadership in decision-making, excellence in practice and innovation, and results that have had a substantial impact at the grassroots level.

Three Priority Themes

In order to meet its objectives, APWF has focused its efforts on three Priority Themes:

- Water Financing and Capacity Development;
- Water-related Disaster Management; and
- Water for Development and Ecosystems.

Progress towards each priority theme will be presented here at the Summit. I would like to set the stage by highlighting the key messages and recommendations from each of the Priority Themes. My speech is based upon the excellent Policy Brief which has been prepared by the whole APWF family, under the supervision of my two Vice-Chairs. I have also drawn inspiration from a new publication by ADB, entitled "Asian Water Development Outlook". I want to thank Mr Kuroda, President of the Asian Development Bank for this important contribution to the Summit.

Water Financing and Capacity Development

Under the First Priority Theme, Water Financing and Capacity Development, APWF members have arrived at the conclusion "*that water problems of developing countries are manageable.*" What this means is that the knowledge base required to evolve proper solutions in the Asia Pacific region is currently available. However, in order to formulate and implement solutions in a timely, cost-effective, pragmatic and optimum manner will require significant additional investment, strong political will, and appropriate capacity development at all levels. The principal challenges are not technological – the hardware of water sector development – but rather the soft issues: sound policies, legal and regulatory reform; planning and technology choices; financing mechanisms; institutional reform and capacity development.

Water and sanitation must be given a higher priority on the national, regional and international agendas – and this requires a firm and continuing commitment from all our decision-makers and political leaders.

A regional investment of just US$8 billion a year would ensure that the Asia-Pacific will meet the MDG Target on water and sanitation. This, in turn, would go a long way towards meeting the combined and complementary objectives of development and poverty reduction, economic growth and social equity in the region.

Water Related Disaster Management

The second Priority Theme, Water-Related Disaster Management, has been seeking to create a new momentum for achieving a positive change in the way water-related disasters are managed in the Asia-Pacific region. As recently highlighted by the IPCC, the increasingly extreme climatic events, such as, floods, droughts and coastal flooding, induced by global warming is likely to further aggravate the impact of disasters in the region. Heavy rainfalls, which are very likely to increase in frequency, will augment the risk of floods. In the course of the next century, the volume of water stored in glaciers and snowcaps, which are melting, are projected to decline, reducing the availability of water that will affect more than one-sixth of the world's population. And, at the other end of the geographic spectrum, the vulnerability of small island states to water shortages, tropical cyclones and sea-level rise continues to increase dramatically.

These climatic projections illustrate the importance of recognizing "adaptation" to increasing risks from climate variability and change as one of the "highest" priority issues. It is, therefore, important to integrate water-related disaster risk reduction into national development plans and to link disaster reduction to national Integrated Water Resources Management plans. Establishing local and national targets for water-related disaster reduction, increasing capacity for monitoring and for early warning systems, and the development of preparedness indices are among the key recommendations that will be presented during this Summit.

Water for Development and Ecosystems

The third and final Priority Theme focuses on Water for Development and Ecosystems. The challenge here is to secure water for long-term, multi-sectoral needs, generating both development and environmental dividends. This includes adopting practical tools to improve the economic, social and environmental outcomes of water management. Water management must recognize that ecosystems themselves require certain levels of water flows in order to remain productive and provide for the needs of all stakeholders in society.

This Priority Theme recommends: (i) the protection and restoration of urban environments, (ii) the modernization and management of irrigation systems and (iii) the adoption of environmental flows in water development planning and policies.

Guiding Principles

The APWF has also developed a series of Guiding Principles to achieve our objectives and these include:

- Empowering the local authorities to which delivery of basic water services has been delegated, and strengthening their capacity to fulfill their role;
- Promoting partnership approaches between local authorities, Non-Governmental Organisations and community-based groups, as well as public and private sector service providers; and
- Providing resources for capacity development to ensure that the knowledge, skills and behavior changes are in place at the levels where these matter most.

Five Key Results Areas

In order to achieve tangible and concrete progress towards each of the priority themes, and towards water and development in general, the APWF also developed five Key Result Areas (KRAs): (i) Developing knowledge and lessons; (ii) Increasing local capacity; (iii) Increasing public outreach; (iv) Monitoring investments and results; and (v) Supporting the work of the Forum and Summit, each of which encompasses a specific set of related activities.

Each of these Key Result Areas has generated a set of Concrete Initiatives. Some of these initiatives are well under way, while others are in the planning stages. However, every single one of these initiatives provides another piece to the puzzle of water security and developmental sustainability. These include the creation of platforms for knowledge coordination, sharing and management as well as monitoring at the regional and country levels.

Impact on People at Grass Roots

All of these messages, recommendations and initiatives are good and commendable. However, our challenge is to implement them. With greater support from leaders and key decision-makers, we can achieve progress on the ground that will make a real difference to the lives of hundreds of millions of people in our region, especially our poor people.

Relationships with Other Major Events

The outcomes of the 1st APWS will feed directly into other high profile international events such as the thirteenth United Nations Climate Change Conference which begins in Bali today, the 2008 G8 Summit in Japan and the 5th World Water Forum. With 2008 earmarked as the International Year of Sanitation, this Summit will help to focus the region's attention on the equally challenging problem of sanitation.

Conclusion

On behalf of the Asia Pacific Water Forum, its lead organizations, and of its many members from throughout the region, I thank you for participating in the 1st Asia Pacific Water Summit and committing to implement its deliverables. Let us agree to embrace our shared vision of raising water and sanitation to the top of our local, national and global agendas. With political will and the support of all the stakeholders, we can make the dream of providing all the peoples of our region, sustainable access to affordable safe drinking water and modern sanitation a reality.

Thank you.

Keynote Speech

His Royal Highness Prince Willem-Alexander of The Netherlands
Chairman of United Nations Secretary-General's Advisory Board on Water and Sanitation

Your Imperial Highness, Prime Minister, Mr. Mori, Your Excellencies, Mr Chairman, distinguished colleagues, ladies and gentlemen.

I am delighted to be with you in Japan, a country that continues to demonstrate a deep commitment to facing and resolving water issues both at home and in the rest of the world.

It is a particular pleasure to be here with His Imperial Highness the Crown Prince of Japan, who has recently agreed to be the Honorary President of the UN Secretary-General's Advisory Board on Water and Sanitation. In doing so, His Imperial Highness proves Japan's invaluable contribution to the Board's work, which was first led by my predecessor, the late former Prime Minister Hashimoto. It was in his honour that we renamed our compendium for action the Hashimoto Action Plan. It is a short – 12-page – action-oriented report, and is the only report UNSGAB will write. I will return to it later.

At this Summit of Asian and Pacific leaders, we are continuing an ancient historical tradition – to discover more effective ways to manage water. Human history is intimately intertwined with water. Social and technological progress can be linked to successful attempts to live with water. Solving water problems in both rural and urban contexts is one of the keys to unlocking economic growth.

I whole heartily welcome and endorse the aim of this Summit to contribute to sustainable water management in order to achieve the MDG targets in Asia and the Pacific.

The Asia-Pacific region faces some of the world's biggest and most difficult water management problems. The recently published fourth assessment report of the IPCC predicts that climate change will cause a number of serious water problems in your region. Freshwater availability will decrease, water security problems will intensify, extra pressure will be put on natural resources and the environment, agricultural and forestry production will decline and sea level rise will threaten coastal areas and especially Small Island States.

Let me quote from the newly published UNDP Human Development Report 2007: "Climate change is the defining human development challenge of the 21st century. Failure to respond to that challenge will stall and then reverse international efforts to reduce poverty. The poorest countries and most vulnerable citizens will suffer earliest and most damaging setbacks, even though they have contributed least to the problem. Looking to the future, no country – however wealthy or powerful – will be immune to the impact of global warming."

Today sees the opening of the UN Climate Change Conference in Bali, which will discuss these very issues. I sincerely hope that our colleagues in Bali will be able to agree on real results and commitments to tackle this human development challenge!

The Asia-Pacific region is also facing enormous challenges in terms of unmet needs in water and sanitation and water-related disasters. According to the Fourth World Water Forum Regional Report, 700 million people living in Asia and the Pacific are without access to clean drinking water, and over 1.9 billion live without access to improved sanitation. Ladies and gentlemen, all speakers have the same figures today, which is unique but at the same time the more frightening to hear the numbers again and again.

The Fourth World Water Forum Regional Report concluded with a call for concrete actions and commitments in relation to financing, water-related disasters, and water for development and ecosystems. I am happy to see that today we are building on those outcomes. For indeed it was last year in Mexico, at that same Forum, that our Board released the Hashimoto Action Plan to advance the world's progress towards the Millennium Development Goal targets on water and sanitation.

The Hashimoto Action Plan identifies six areas most in need of focused, coordinated action: Financing, Water Operators Partnerships, Sanitation, Monitoring, Integrated Water Resources Management, and Water and Disaster.

Those six areas of action and this Summit's Priority Themes have much in common.

On financing, our Board believes that the central issue is to divert international monetary flows into water and sanitation sectors. Capacity building and local financial schemes adapted to water must be promoted not separately but in concert. The Hashimoto Action Plan accordingly identifies three key actions. First, we must create better governance and transparency in water services. Second, we must expand local service operators' knowledge and awareness of new funding sources while also developing local financial markets. Third, donors should devote their water funding to building capacity in these areas. In the last year and a half, our Board has signed agreements furthering these principles with the OECD, the Asian Development Bank, the African Development Bank, and the Inter-American Development Bank.

The Hashimoto Action Plan notes that since public water services currently provide more than 90% of the world's water supply, improvements in their capacity would have an immense positive impact on the provision of water services globally. Accordingly, we have proposed a new mechanism to improve public water services based on the concept of mutual support through twinning agreements between water operators, public or private, on a not-for-profit basis. We call this structured programme of cooperation Water Operators Partnerships, or WOPs for short. We were delighted when UN-HABITAT agreed to form the Global Water Operators Partnership Alliance. Already, WOPs are flourishing in Asia, with the Asian Development Bank and the Global Water Partnership collaborating to support Asian water utility operators.

Water is life. Water in the form of flood, tsunami or mudflow is also a threat to life. To lower the human and social costs of such tragedies, the Hashimoto Action Plan suggests that the international community must agree on targets and goals for coping with water-related disasters to create global awareness and political commitment. When disaster comes, water is life again: the displaced need adequate water and sanitation in their temporary shelters. We are working with relief agencies to devise a modular repeatable approach to this challenge. To facilitate this, a High-Level Expert Panel on Water and Disaster met this September in Tokyo. It formulated objectives for funding, developing guidelines for integrating information on flood projections. Cyclone Sidr, which struck Bangladesh with relentless force a few weeks ago, is a good example of what can be achieved. Although every human life lost is one too many, the difference between half a million in the early 1970s and several thousands in 2007 is striking. The solutions are not high-tech or only available to the developed world. I look forward to exploring synergies between this expert panel's work and our discussions on water and disaster over the next few days.

Next year will be a pivotal year for water and sanitation issues. The Hashimoto Action Plan recognised the acute need to galvanise political will to address the sanitation crisis. And so we proposed an International Year. We were delighted when the UN General Assembly unanimously declared 2008 the International Year of Sanitation. It was officially launched on 21 November in New York.

I look forward to the regional launch tomorrow and the opportunity to think hard about how best to capitalise on this opportunity to bring improved sanitation to the people of this region. Sanitation is the foundation for health, dignity and development and next year we must make progress.

In 2008, Japan will host the G-8 Summit, and UNSGAB is calling for the G-8 members to re-examine "Water – a G-8 Action Plan," which they adopted in Evian five years ago. Next year, we believe, is pivotal, and we think the G-8 could make an important contribution by renewing their commitment to water and sanitation. Preferably not as item number x under the popular topic of climate change, but as a specific issue, with the full attention deserved by the 2.6 billion of our fellow human beings who still live without adequate sanitation.

Revisiting progress made on agreed targets is another important area of the Hashimoto Action Plan. We must have effective data collection, monitoring and reporting tools so we can determine if progress is being made. We are currently looking at how much to strengthen the Joint Monitoring Programme by WHO and UNICEF, which monitors progress in water delivery and sanitation.

Integrated Water Resources Management is an internationally acknowledged approach that seeks to avoid the lives lost, the money wasted, and the natural capital depleted because of decision-making that did not take into account the wider ramifications of sectoral actions. Our Board believes IWRM is a flexible tool for optimising water contribution to sustainable development, while also ensuring water for ecosystems and environmental flows. At the World Summit on Sustainable Development in Johannesburg, countries agreed to develop national IWRM Plans and the Hashimoto Action Plan suggested that countries should share their experiences at the sixteenth session of the Commission on Sustainable Development in May 2008. The United Nations responded, and requested that countries share their Plans for review at CSD-16. These Plans are currently being collected and analysed by the Department of

Economic and Social Affairs and we look forward to a substantive discussion on IWRM next year.

Our Board will be very encouraged if Asian and Pacific leaders take the initiative to develop a regional Ministerial Council on water and sanitation. We commend the effort already made and we urge bold steps to institutionalise the progress so far. In Africa, the African Ministers' Council on Water has been invaluable in providing political leadership and advocacy for water and sanitation. It has, for example, been directly involved with our Board in promoting an African Union Summit on Water and Sanitation next year. A similar model here in the Asia-Pacific region could make a significant contribution to advancing progress towards the Millennium Development Goals.

Ladies and gentlemen,

The Asia-Pacific region is facing enormous water challenges. Climate change, river basin management, drinking water and sanitation and water-related disasters all demand a swift, appropriate response. Concrete measures are needed to bring the MDGs within reach and to assure sustainable development for all countries and the chance of a safe and healthy future for all your citizens.

Your presence here testifies that you have the will to cooperate and to find solutions. I can assure you of UNSGAB's full support in this difficult process. And on behalf of our Board, I wish you all an interesting, insightful and very productive Summit.

Thank you.

Message

H.E. Mr. Ban Ki-moon
United Nations Secretary-General

Your Highness, Crown Prince Naruhito, Your Highness, Crown Prince Willem-Alexander, Prime Minister Yasuo Fukuda, Former Prime Minister Yoshiro Mori and Distinguished Delegates.

Our planet faces a growing water crisis. But the situation in the Asia-Pacific region is especially troubling. High population growth, unsustainable consumption, pollution and poor management all threaten the area's clean water sources.

Climate change is making a bad situation worse. Glaciers are receding. Floods are getting worse. Droughts are becoming longer and more severe.

Despite these warning signs, water care remains an underserved and underappreciated field. Throughout the world, water resources continue to be spoilt, wasted and degraded.

The consequences for humanity are grave. Water scarcity threatens economic and social gains. It undermines environmental sustainability. It slows progress towards the Millennium Development Goals. And it is a potent fuel for wars and conflict.

This crisis demands leadership. This inaugural Asia-Pacific Water Summit is an opportunity to answer this call. I commend all of you for committing your time and energies to it. And I thank former Prime Minister Mr. Yoshiro Mori for leading the initiative.

Your work today promises big returns tomorrow. Investments in water and sanitation are among the best we can make: for every dollar spent, we can expect to save at least three dollars, and as much as 34 dollars, on health, education, and social and economic development.

Of course, the benefits of clean water cannot be measured in dollars and cents alone. There is no price-tag on transforming lives, or on giving every child a real chance for a healthy, productive future.

I hope you will use this Summit to seek practical solutions to pressing water problems. And I hope you will work to make the Asia-Pacific region a true leader in this field.

Thank you very much.

Commemorative Lecture

His Imperial Highness the Crown Prince of Japan

Humans and Water

- From Japan to the Asia-Pacific Region -

1. Introduction

I am very pleased to have the opportunity to deliver this commemorative lecture at the 1st Asia-Pacific Water Summit.

First of all, I would like you to look at this photograph (Figure 1). I took this photo when I visited Pokhara in Nepal in 1987. Women and children have come to collect water. As you can see, water barely trickles from the tap. I recall thinking to myself, "My goodness! How long will it take them to collect enough water? And most of them are women and children. What hard lives they lead!"

Then I learned that the status of women has not improved in many developing countries even today. They have still not been freed from the heavy chore of fetching water. As for the children, they spend their time collecting water instead of going to school. This is the stark reality for so many of the poor of the world. I also learned that most issues stemming from global warming significantly affect the entire global ecosystem as well as human society through the water cycle. These realizations made me deepen my interest in water, which I realized, has implications that reach far beyond my original field of study regarding water transport. In fact, water affects our lives and societies in many more ways, not only in terms of water supply and flood control but also in regard to sanitation, environment and education.

Photo by the Crown Prince Naruhito

Figure 1 Women and children gathering to collect water -On the way to Sarankot Hill-

It was with this concern that I happily accepted to serve as the Honorary President of the 3rd World Water Forum in Kyoto, Shiga and Osaka in 2003. I spoke then on "Waterways Connecting Kyoto and Local Regions", describing how water transport served the former capital city of Japan. At the 4th World Water Forum that was held last year in Mexico City, I spoke on how water had played a profound and extensive role in serving the development of Tokyo when it was known as Edo.

This time, I would like to share with you my own thoughts about water, starting from water transport on the Seto Inland Sea, which includes Oita Prefecture, by referring to the history of the relationship between water and the Japanese people, as well as more broadly to global water issues.

2. Water supporting exchange

(1) Ancient and Medieval Water Transport on the Seto Inland Sea

Oita prefecture, where this Summit is taking place, lies in Northern Kyushu. From ancient times and up until the Middle Ages, the two major centers of Japan were

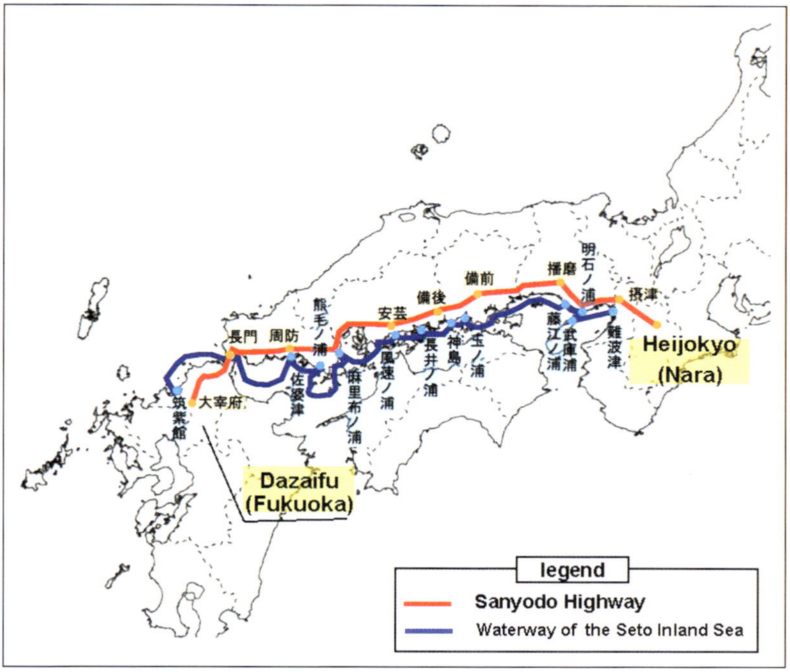

Map provided by the Japan Water Forum

Figure 2 Sanyodo Highway and Waterway of the Seto Inland Sea

Kinai, now the Kinki Region, where the capital such as Nara and Kyoto was located, and Northern Kyushu. Land transportation between them was conducted along the Sanyodo highway (Figure 2). There were seven highways that extended from Kinai to the rest of the country, but the Sanyodo highway and its extension to Dazaifu was the only highway designated as the "Daiji" or trunk road. Dazaifu not only served as the branch office of the government in Northern Kyushu but also was the base for foreign and military relations. The ancient Yamato dynasty always recognized the importance of Dazaifu.

The Seto Inland Sea connected the two major centers as a waterway for people and goods. Here in Oita, Kunisaki no tsu, which is thought to be located in Kunisaki City, and Sakato no tsu, which is thought to be located in Oita City, are known to have been major ports along this waterway from ancient times.

An interesting chronicle of transport and trade in the Seto Inland Sea has been preserved in a document dating back to 1445 titled 'Hyogo-Kitaseki Irifune Nocho', or 'Registration records of ships at the North Customs-post of old Hyogo Port'. Hyogo port which is called Kobe today was then owned by Nara's Todaiji Temple.

The document, which consists of customs records collected from ships that entered the North Customs-post of Hyogo at that time, is considered an important historical document comparable to customs records kept in the mid-fourteenth century in Lubeck in northern Germany.

The record tells what goods were shipped from which port and how much customs duty was levied. The annual number of ships entering the North Customs-post of Hyogo was about 1,960, suggesting the prosperity of the port and its region. The main commodities traded were salt, rice and lumber-18,000 cubic meters, 4,500 cubic meters, and 11,000 cubic meters respectively.

The preeminence of salt reflects the natural characteristic of the Seto Inland Sea region, which with its abundant sunshine and little precipitation was conducive to salt production. It also speaks of the wide use of salt to preserve food, and further, that there was a large concentration of population in Kyoto and the surrounding areas. As for rice, one's eye is caught by an entry for "red rice" though in relatively small quantity. From the ship's registry it is assumed to have been produced in Sanuki, currently Kagawa Prefecture, which has suffered shortage of water from ancient times. It tells us of another blessing of exchange through waterways that drought-tolerant red rice originating in Southeast Asia was grown in Sanuki.

(2) Exchange with Asia and the world

Kinai was not only linked with Northern Kyushu, as we have seen, but also had links to Asia and the rest of the world through the Seto Inland Sea. In ancient times, envoys were sent to Tang China (Figure 3), and during the Muromachi Shogunate (14th to 16th century) Hyogo Port was the center of trade with Ming China. In addition to trade with China, trade with the Yi dynasty of Korea prospered and cotton was imported into Japan for the first time. This not only brought an enormous change to the clothing habits of the Japanese who until then wore only silk and hemp, but eventually encouraged the growth of a new industry, cultivating indigo for dyeing the material. Tokushima, in particular, which was located in the Seto Inland Sea Region, became known for the production of indigo for the whole country.

In the 16th century, we know that the Jesuit missionary Francisco Xavier reached Sakai and Kyoto via the Seto Inland Sea. During the period of Sorin Otomo, a general who ruled Northern Kyushu at that time, Oita was one of the bases for Christian missionary work. Francisco Xavier is known to have visited there (Figure 4).

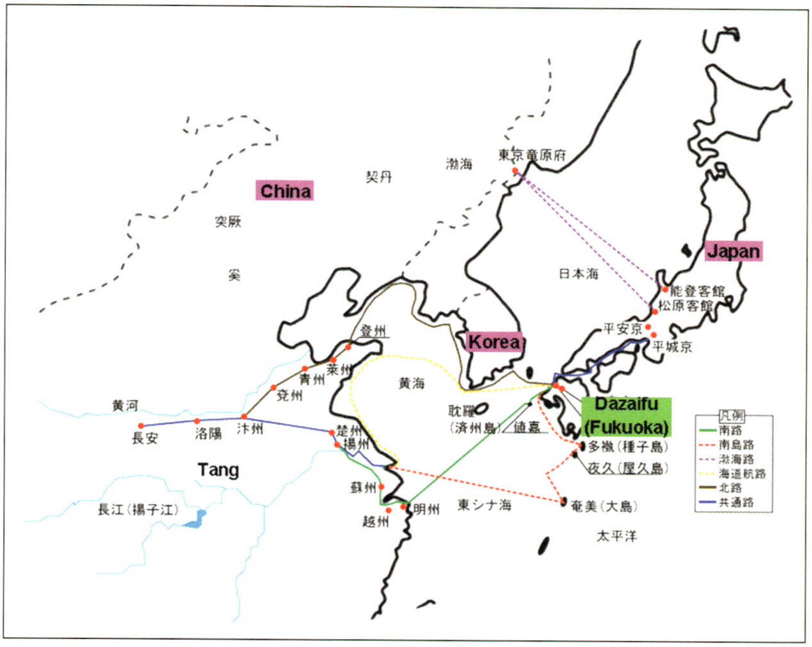

Map provided by the Japan Water Forum

Figure 3 Maritime route to Tang China

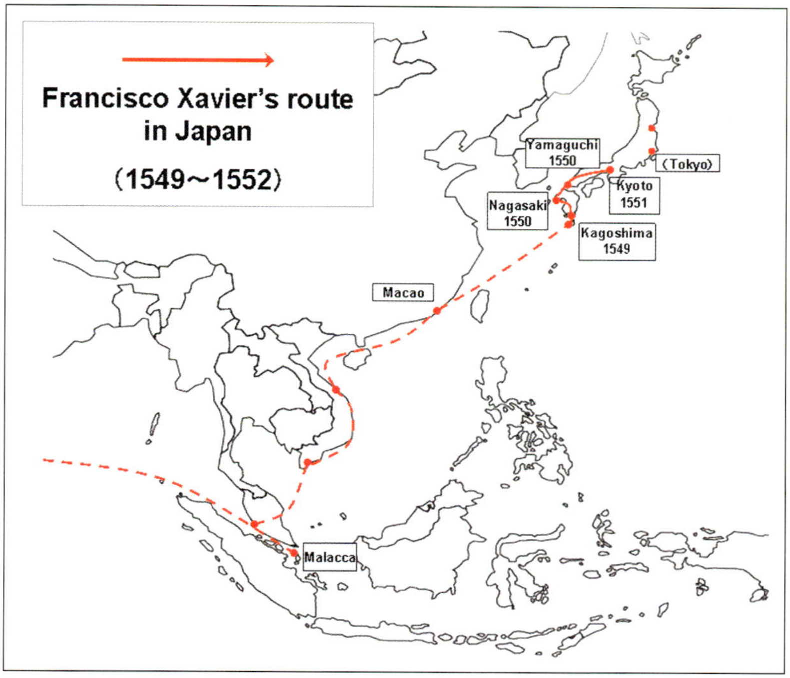

Map provided by the Japan Water Forum

Figure 4 Route that brought Francisco Xavier to Japan

Nowadays, when transportation is predominantly by air, rail and road, we are apt to look upon the seas and rivers as barriers separating countries and regions. Historically, however, one can say that not only countries and regions but also people were linked by the seas and rivers through water transport.

3. Diversity of water's nature and role

(1) Blessings of water

Water has many other roles to play in addition to that of facilitating trade and transportation. Humans cannot live without water, and needless to say water has played a vital role in human history.

I would like you to look at this photograph (Figure 5). This is a satellite photo of Ulan Bator, the capital of Mongolia. One sees that the city has developed along tributaries of Tuul River that runs in the south of the city.

I took this picture in the ancient city of Qaraqorum (Kharkhorin), when I visited Mongolia in August this year (Figure 6). The city was built beside Orkhon River.

METI and NASA 2007
all rights reserved, processed
and distributed by ERSDAC, Japan

Figure 5 Satellite photo of Ulan Bator

Photo by the Crown Prince Naruhito

Figure 6 Near the ancient city of Qaraqorum

The little rain this year explains the dry-looking scenery in the picture, which usually looks greener. I had imagined Mongolia to be a country of steppes and desert, and I was struck to see that much as in other regions of Asia, people here too built cities along the rivers to take advantage of their blessings.

One does not have to study very deeply the world's four major early civilizations to observe that people have been able to progress with the many benefits of water. In every period and in every region, however, they could not enjoy only the blessings of water. Our history has been marked by battles for water and with water-ours and that of other peoples, too little or too much.

(2) "Shortage of water" and "Surplus of water"

Let us return to the example of the Seto Inland Sea to examine the history of our battle with water; first, its shortage. This photograph is that of Tashibu Manor that once belonged to Usa Shrine (Figure 7). The terraced paddy fields have been created by the efforts of generations of people. Drawing water from the springs on the premises of Amabiki Shrine, the water is designed to flow down from the paddies at the higher levels to those below through gently curved conduits. This beautiful landscape, one may say is the fruits of the wisdom and sweat of many generations of people.

Photo credit: Bungotakada City

Figure 7 Tashibu Manor (Bungotakada City, Oita Prefecture)

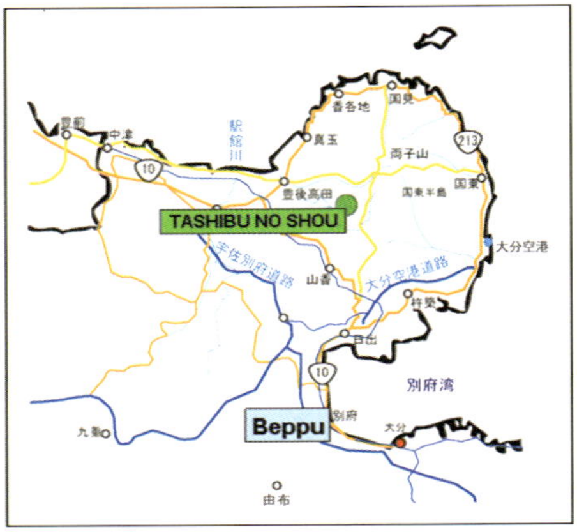

Location of Tashibu Manor

Fukiji Temple

Kumano Magaibutsu Stone Buddha

Photo credit: Bungotakada City

Figure 8

Fortunately, the landscape of the manor is preserved for us to see. Also the whole area is preserved as an attractive historical site with representative cultural assets of Kunisaki Peninsula including Fukiji Temple, Maki Odo Temple and Kumano Magaibutsu Stone Buddha (Figure 8).

People did not make these terraced paddy fields for the sake of a scenic view. This is the result of their efforts to most efficiently make use of the natural resources. It must not be easy to maintain the terraced paddies under these geographical conditions. However, the well maintained terraces not only brought harvests but also in times of heavy rain served as reservoirs to reduce the menace of flooding the plains below. The water drawn into the paddies penetrated the soil to enrich the downstream area's groundwater. Such human efforts resulted in a small new water cycle.

Please look at this figure (Figure 9). It shows the distribution of reservoirs in the Seto Inland Sea Region. 60 percent of reservoirs in Japan are concentrated in this region. This is a photograph of Manno Pond, an irrigation pond in the Sanuki plains

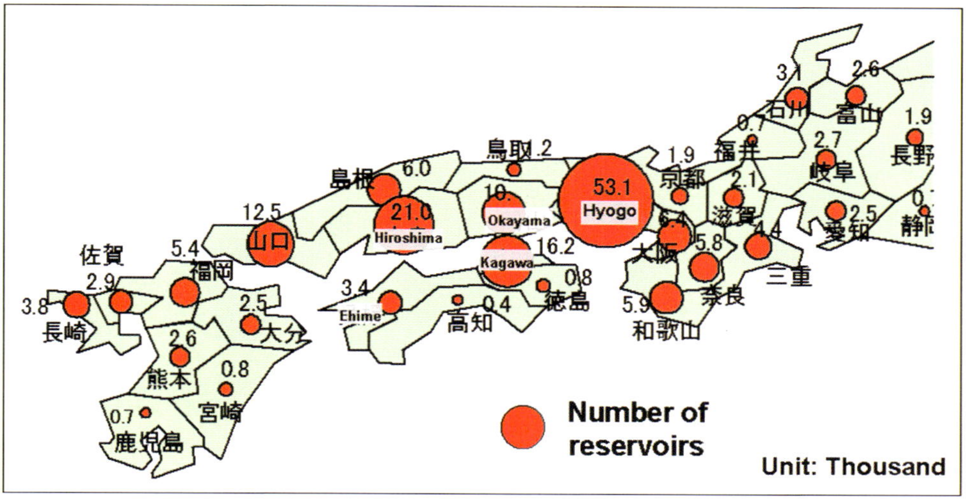

Source: The Ministry of Agriculture, Forestry and Fisheries of Japan, Inventry of Ponds in 1991

Figure 9 Distribution of Reservoirs in the Seto Inland Sea Region

Photo credit: Kagawa Prefecture

Figure 10 Manno Pond (Manno Town, Kagawa Prefecture)

(Figure 10) said to have been restored by Kukai who is known respectfully by his posthumous name "Kobo Daishi", the reverend Buddhist master. Since the pond was made in the early 8th century it is still a precious source of water in the Sanuki Plain which has suffered from the shortage of water.

Photo credit: Osaka Prefecture

Figure 11 Sayama Pond (Osakasayama City, Osaka Prefecture)

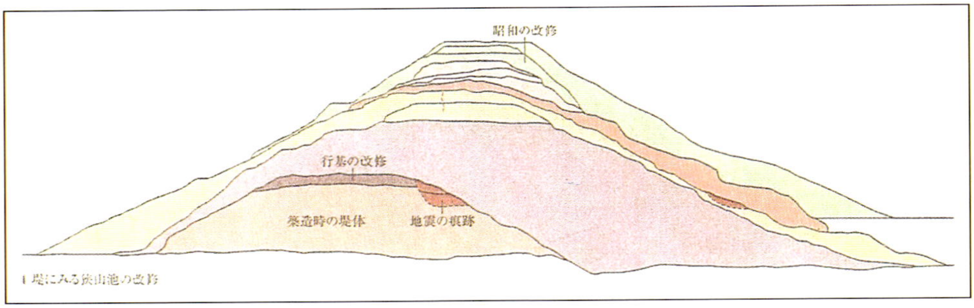

Copyright: Osaka Prefectural Sayamaike Museum

Figure 12 Figure of historical cross section of the embankment of Sayama Pond

This is Sayama Pond in Osaka Prefecture. It is said to have been built by Gyoki, a great Buddhist monk (Figure 11). It is also still in service, and the latest restoration work took over twenty years from 1980 to 2001. At an excavation in the restoration work, it was revealed that it can be traced back at least to the year 616, a hundred years before Gyoki's period. The survey revealed that its bank had been extended a number of times to reach what it is today (Figure 12). Gyoki is thought to have taken part in some of the restoration.

A 'Shikiha' method of bank building used in ancient China and Korea was also used in building and restoring the banks of Sayama Pond. This technology was brought to Japan by exchange through water transport as well. Here again we see the benefits of exchange mediated by sea.

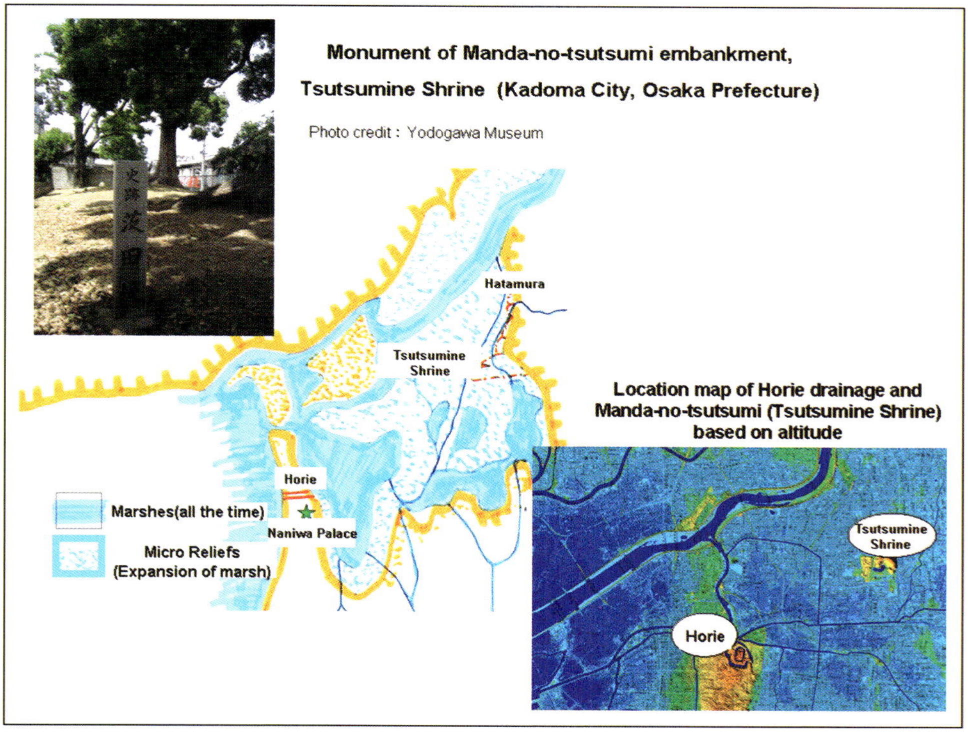

Photo credit: Yodogawa Museum

Figure 13 Horie drainage and Manda-no-tsutsumi Embankment

In this way, we are managing the issue of the "water shortage" of the respective regions by continuing to benefit from the efforts of our forefathers and adding our efforts to improve on them.

Let us now turn to the issue of "surplus water".

The oldest record regarding surplus water in Japanese history is seen in Nihon Shoki (Chronicles of Japan) written in the 8th century. We can read there that in the reign of Emperor Nintoku, Manda-no-tsutsumi embankment and Horie drainage were built to prevent flooding of Yodo River and to develop the northern Osaka Plain (Figure 13). Based on the research done on the ancient flood control and water utilization, it can be assumed that it was built in the early 5th century. Even today, Horie flows through the center of Osaka as the old Yodo River (Okawa River) and Manda-no-tsutsumi embankment retains its form in the vicinity of Osaka City, such as Neyagawa City and Kadoma City. I am amazed by the scale of imagination of our forefathers in working on two projects at once, the drain and the bank, which are 10 kilometers apart.

Land is essential for settlement and production, but it is usable only if humans provide the necessary water and manage its excess. It was said in ancient China as well as in Japan that "those who govern water govern the country". What is meant by governing water does not mean single projects such as ensuring the supply of water and controlling floods. It should be taken to mean, instead, the broad task of managing nation building. Water Governance, I believe, is exactly what our forefathers have done, to develop our national land, by continuing to manage water comprehensively.

(3) Japanese people and water in various relations

Beyond the Seto Inland Sea Region we see how Japanese people have related to water. As mountainous areas cover four-fifth of the national land, they have an enormous impact on the water cycle of each region.

This is a photograph of Mount Chokai, which is located in the northeastern region of Honshu (Figure 14.1, 14.2, 14.3). I was told when I climbed the mountain last

Photo by the Crown Prince Naruhito

Figure 14.1 Mount Chokai (View of Mount Chokai from Ryugahara Marsh)

Photo by the Crown Prince Naruhito

Figure 14.2 Mount Chokai (Snow patch)

Photo by the Crown Prince Naruhito

Figure 14.3 Mount Chokai (Shishigahana Marsh, Nikaho City, Akita Prefecture)

Photo by the Crown Prince Naruhito

Figure 15.1 Mount Aso - Nakadake and Eboshidake - (Nakadake)

Photo by the Crown Prince Naruhito

Figure 15.2 Mount Aso - Nakadake and Eboshidake - (Eboshidake)

year that the snow on the mountain melts and nourishes the beech trees, and its underflow moistens the fields at the foot of the mountain, and eventually it flows into the Sea of Japan and nourishes rock oysters. Rain and snow fill the river and percolate underground changing its form to nourish nature and support lives of the people engaged in agriculture and fishery. And there are those who contribute to managing and protecting water. Here again we can see how we relate to the water cycle.

In Kumamoto Prefecture, which lies to the west of Oita Prececture, forest covers 60 percent of its land and it nurtures the quality and quantity of groundwater. 80 percent of the water supply comes from groundwater. The entire water supply of Kumamoto city, known as "The Water City", depends on underground water stored within the layer of pyroclastic flow deposit from Mount Aso (Figure 15.1, 15.2). We can see how the Japanese have related to water based on various natural conditions.

Water also deeply relates to our hearts and minds. At Todaiji Temple in Nara, our ancient capital, there is an event known as Shunie (Figure 16). Omizu-tori, water drawing, is a mystic ritual of drawing water from Wakasa Well at midnight on the 12th of March every year, which, according to the legend, comes from faraway Wakasa Bay. The ritual is said to relate to a folk custom supposed to ward off evil by drawing water before the arrival of spring. Shunie is also an epic festival of fire in which huge torches are brought into the hall, carried through the balcony and crushed on the floor. This festival of water and fire which combines the two most fundamental elements of our existence has been celebrated every year for the last 1250 years without a single interruption. This may be the result of devout feelings in the depth of people's hearts.

Otaimatsu

Omizu-tori

Photo credit: Mr. Eisuke Ueda

Figure 16 Shunie (Omizu-tori), Todaiji Temple

4. Global water issues: present and future

(1) Millennium Development Goals

Let us now turn to global water issues. If we look at the UN Millennium Development Goals which sets eight goals to be achieved by the international community by 2015, it is not an exaggeration to say that water is deeply related with each of them. For example, if we are to raise the status of women and achieve universal primary education, women and children must be liberated from water-fetching chores and be freed from the mortality caused unsanitary water-borne diseases such as cholera and typhoid fever. I am seriously concerned by the fact that one child dies every 10 to 20 seconds due to water related diseases.

Let us look at the availability of safe drinking water and the provision of basic sanitation. While there is fairly steady progress in achieving a sufficient supply of water, there is a serious delay in meeting sanitation. We can see the seriousness of conditions especially in the rural sector. This is true for the Asia-Pacific region as well (Figure 17, 18, 19).

(2) The International Year of Sanitation

As you know, next year is the International Year of Sanitation. It is hoped that adequate efforts and public education programs will be pursued in solving the water and sanitation problems which lag seriously behind other efforts. In order to promote

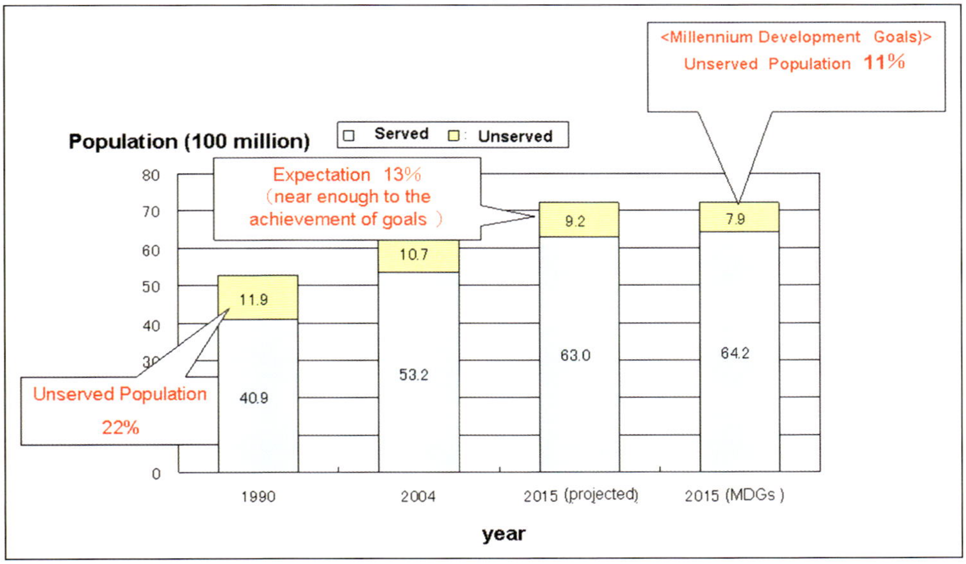

Source: WHO/UNICEF JMP 2006

Figure 17 World population without access to an improved drinking water source

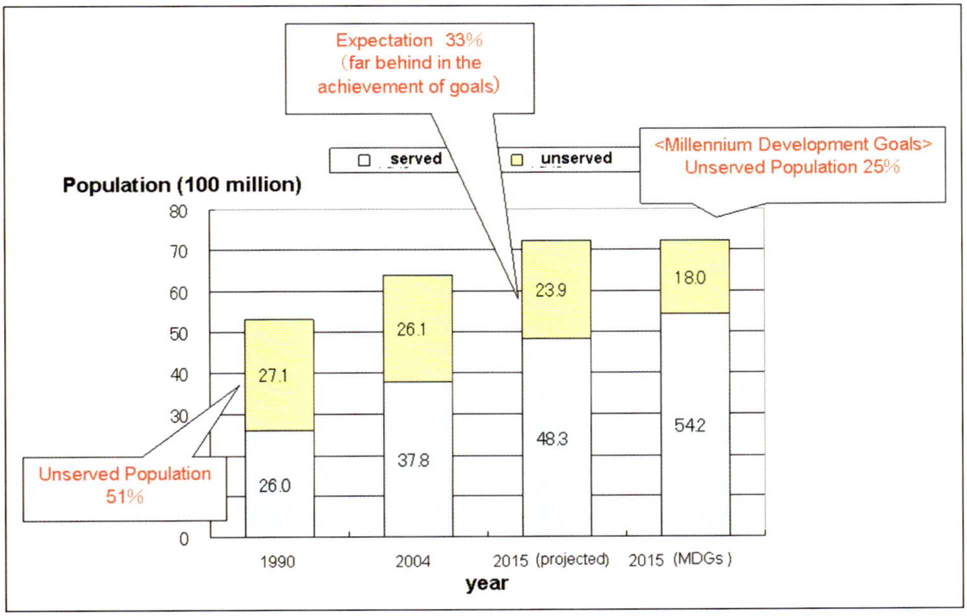

Source: WHO/UNICEF JMP 2006

Figure 18 World population without access to improved sanitation

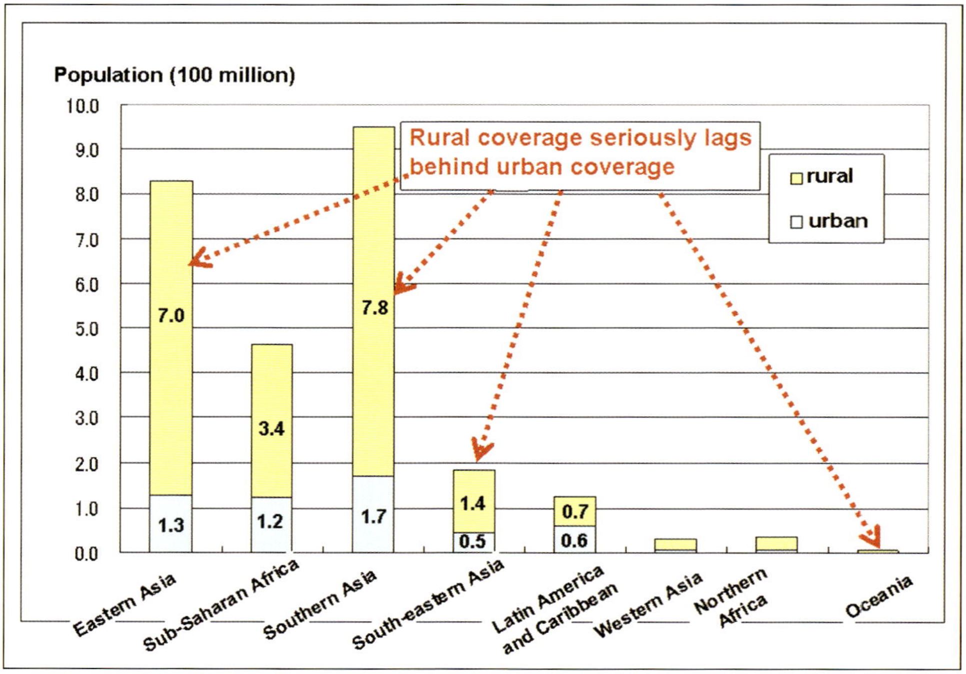

Source: WHO/UNICEF JMP 2006

Figure 19 Urban and rural population without access to improved sanitation

these efforts, not only is the cooperation between countries important, but also the construction of networks between local governments that are responsible for the water supply and sanitation facilities, and civil organizations which work in the field, are necessary. Some actions have already been taken, which are expected to be emphasized by the International Year of Sanitation.

Furthermore, we may have to adopt new approaches with regard to the provision of basic sanitation facilities themselves. For example, I understand that there is an increased focus on ecological sanitation-ways to minimize the use of water and reuse human waste as fertilizer safely and effectively. This is a "bio-toilet" installed in cabins on Mount Fuji (Figure 20). I am told that such facilities have been installed in our conference site for participants to test. Although there is still more work to be done to reduce their cost, in order for them to be widely used in the developing countries, this is certainly a technological innovation with great chance of success.

(3) Global Warming

Water is very closely related to the climate change issues, which are feared to seriously threaten the very survival of the human species. Global warming is predicted to cause

Cabin at the eighth station/Fujinomiya-guchi Trail
toilets 6, urinals 4, capability 360 times/day (Sawdust bio-toilets)

Cabin at the seventh station/Subashiri-guchi Trail
toilets 5, urinals 2, capability 360 times/day (Sawdust bio-toilets and Incinerating-toilets)

Photo credit: Shizuoka Prefecture

Figure 20 Bio-toilets on Mount Fuji

not only a rising sea level and greater frequency of abnormal weather patterns, but also the intensity of natural disasters and widespread droughts. In recent years, there have been increasing instances of disastrous flooding in many parts of the world, while a part of other areas suffer from worsening drought. I am deeply saddened by the enormous damage and human tragedy brought by water-related disasters in the Asia-Pacific region.

According to the IPCC 4th assessment report issued this year, serious repercussions in the Asia-Pacific region such as increased floods and snow avalanches caused by the melting of Himalayan glaciers are predicted (Figure 21). Many leaders from the Pacific area are here with us. It is feared that the island nations of the Pacific may suffer from coastal disasters such as submergence, tidal waves and erosion due to a rise of the sea level, as well as from a decreasing availability of fresh water supplies. I believe that these are important themes to be discussed in this Summit and hope that fruitful discussions will be to solve these problems (Figure 22).

The Himalaya (Eastern Nepal) in 1978 The Himalaya (Eastern Nepal) in 1998

Melting of Himalayan glacier is expected to increase floods, snow and rock slides, and bad effect on water resources.

More than one billion people can be affected by water shortage by 2050.

Source: IPCC 4th Assessment Report

Photo credit: Laboratory of Cryosphere Variation, Nagoya University

Figure 21 Melting of Himalayan glaciers

Occurrence of coastal erosion in Tuvalu

Sea-level rise is expected to exacerbate inundation, storm surge, erosion and other coastal hazards, thus threatening vital infrastructure, settlements and facilities that support the livelihood of island communities.

Source: IPCC 4th Assessment Report

Photo credit: Ministry of the Environment of Japan

Figure 22 Impacts of sea-level rise on the small islands in the Pacific

5. Conclusions

I have shared with you some of my thoughts about water, such as how water supports commerce, its diversity of nature and roles, and global water issues. I hope that I was properly able to explain our experience in Japan, how we have made the best use of our terrain and its natural endowments, acquiring knowledge and experience from overseas, and striving to improve the enriching and life-giving connection between people and water.

All water issues are linked. Issues such as water supply, sanitation and flood control do not exist independently. For their resolution, we must understand their diverse nature and broadest implications, and then make continued efforts to find the best measures applicable in local conditions, by not only taking comprehensive and integrative approaches, but also by implementing them in innovative partnerships with all concerned.

I would like to conclude my presentation with a sincere wish that in its deliberations this Summit will take into consideration the rich diversity of the Asia-Pacific region and be forthcoming with proposals for the resolution of common regional issues concerning water, and thus of the global problems of which they are an inseparable part and that affect us all.

Thank you very much for your kind attention.

UNOFFICIAL TRANSLATION

SPEECHES BY HEADS OF GOVERNMENT

- H.E. Mr. Emomali Rahmon, President of the Republic of Tajikistan (Vice Chair)
- H.E. Mr. Anote Tong, President of the Republic of Kiribati (Vice Chair)
- H.E. Lyonpo Dr. Kinzang Dorji, Prime Minister of the Kingdom of Bhutan (Vice Chair)
- H.E. Mr. Tommy Remengesau Jr., President of the Republic of Palau
- H.E. Mr. Ludwig Scotty, President of the Republic of Nauru
- H.E. Mr. Emanuel Mori, President of the Federated States of Micronesia
- H.E. Mr. Mititaiagimene Vivian, Prime Minister of Niue
- H.E. Mr. Apisai Ielemia, Prime Minister of Tuvalu
- H.E. Mr. Dosbol Nur ulu, Vice Prime Minister of the Kyrgyz Republic
- Concluding Remarks of Speeches by the Heads of Government

H.E. Mr. Emomali Rahmon
President of the Republic of Tajikistan

Excellency Chairman, Your Royal Highness, Excellencies, Distinguished participants of the Summit, Dear Ladies and Gentlemen.

As the President of the Republic of Tajikistan and the President of the International Fund for saving the Aral Sea it gives me great pleasure to deliver a speech from a high tribune of this important water summit.

It is especially pleasant being here, in the friendly land of Japan where the 3rd World Water Forum was held in 2003 and our initiative on declaration of International decade of «Water for life» had been presented on that Forum. It was the initiative of the necessity to make the countries to take responsibilities for condition and use of water resources and responsibility for life in the planet.

It is a notable, that this initiative was supported by 141 states including the countries of Asia-Pacific region.

Appearance of new partnership network – Asia-Pacific Water Forum says about an idea of International Year of Freshwater (2003) and International Decade of «Water for life» had found a response in such huge region of the world. Convening this Summit – is an illustration that attention to water, and its irreplaceable essence is continually increasing. World progress tendencies are clearly illustrating that its value can exceed a price for oil, gas, coal and other resources which are necessary for sustainable development of future life of every country and region.

Unfortunately, the facts are saying about nearly 4 thousand children are dying every day because of poor quality water, and approximately more than a billion people do not have an access to freshwater and majority of them are living in our continent. These problems are subjected for corresponding to poor countries that have insignificant potential for their solution. We therefore consider it very important to establish the Regional and International Special Funds for elimination of difficulties concerned with access to freshwater and assisting in water disaster emergency cases under the structure of the United Nations.

In my view, the priorities of Asia-Pacific Water Forum are more clearly reflect a necessity of joint solving of water issues through improvement of water financing, improvement of water factor disasters management and reasonable ecosystem approach.

Water sector financing which is the main prerequisite for achieving one of the important Millennium Development Goals will twice decrease a number of the people who do not have access to secured drinking water and at whole it depends on advanced economy and political will of the leaders of the states. I therefore offer to establish Water Partnership of Developed Countries which would directly assist developing and poor countries in improving water sector and mentioned tasks.

In this regard, in our opinion, helping poor and underprivileged countries should be a primary task of second part of the International Water Decade.

All these issues and proposals require a complex consideration and decision, which arising the necessity of efforts consolidation taken on regional and sub-regional levels. I deem that convening Special Event of High Officials Level within the framework of the UN General Assembly, dedicated to discussion of condition of water issues on a global arena and consideration of concrete steps upon eradication of problems concerned with water would meet this aspiration. Elaboration and adoption of International Water Convention could be one of the most important steps in a unification of efforts which would determine universal principles of water policy ensuring the interests of all consumers.

According its economic growth Tajikistan is gradually increasing the volume of water sector financing. Priority significance is giving to hydro-energy which the hugest resources in this direction can ensure a cheap and ecologically safe electricity not only the economy of Tajikistan but and other neighboring countries as well. We appeal all interested parties for cooperation in assimilation of these resources on the basis of equitable and mutually benefit terms.

Tajikistan which 93% of its territory is difficult relief mountains is more subjected to water related natural calamities. Flood and draught, torrent and freshet, avalanches and landslides are annually causing significant damage to the economy of the country.

A danger of Sarez Lake burst with a volume of more than 17 cubic kilometers located in Tajikistan is threatening a life of 6 million people of Tajikistan, Afghanistan, Uzbekistan and Turkmenistan. I have already proposed to use the cleanest freshwater of Sarez Lake for water-supply of millions people in Central Asia with the establishment of International Consortium.

Support of this idea will allow not only neutralize this risk but also to use this object in the favor of the peoples' interests as well.

Global warming is causing a serious anxiety which in consequence of this phenomenon intensive thawing of Tajikistan high-mountains glaciers is observing.

It is fraught with a change of hydrological regime of the rivers which has disastrous effects not only for our country, but also for downstream states whose irrigative land development is tremendously depends on these resources.

Taking into consideration increasing cases of water-related natural disasters, the Government of the Republic of Tajikistan decided to convene The International Conference on Water-Related Disasters Reduction in Dushanbe next year. We invite the Governments of Asia-Pacific continent, UN Institutions, International organizations, non-governmental organizations, scientists and experts and all who are concerned about this urgent issue to participate at this very important event.

Distinguished participants of the Summit,

A tragedy of Aral Sea is known to world community for a long time. I would like to draw your attention to this global problem once again. In the conditions of global warming and population upsurge this problem has becoming an issue of the day of the Central Asian region. Countries of the region are not able to withstand this challenge. Taking this issue into account, heads of the states of the Central Asia propounded the initiative on giving a status of UN Institute to International Fund for saving the Aral Sea for coordination of international organizations' and donors' activities and declaring Aral Sea basin as a priority field trial region for the sake of meeting Millennium Development Goals.

I am hopeful, that this initiative will find its reflection in the summary documents of the Summit and thereby will promote a solution of majority issues of the Aral Sea Basin countries.

Asia-Pacific Water Forum is being one of the important water communities of Asia and Pacific and in the object of intensification of its role and improvement of water resources management we consider it expedient to set centers in all five its sub-regions. I propose that it is a necessary for new-established Water Forum to concentrate more significant attention to water problems of the Central Asian sub-region. Tajikistan is ready to comprehensively assist a Forum in this direction and offers to place its Central Asian Centre in Dushanbe.

Policy plays a key role in improving water sector. Namely political will of the leaders of the states is one of the main elements in achievement sustainable water development. It is pleasant, that this matter has been gradually coming known all entire the world and political leaders paying their attention to the solving water issues more often. It is remarkable that the results of our Summit will be presented to «G8» Summit which will be convened here on friendly land of Japan next year.

We together have a splendid opportunity to be concentrated on concrete issues of water development in its all aspects within the definite countries, sub-regions and whole Asia-Pacific region, as well as to contribute for approval of progressive sustainable water resources management for the welfare of the people.

I wish all of us every success.

Thank you for attention.

H.E. Mr. Anote Tong
President of the Republic of Kiribati

Mr. Chairman, Our host – the Government and People of Japan, Excellencies, Distinguished delegates, Ladies and gentlemen.

I wish to take this opportunity to express my sincere gratitude to our host, the Government and People of Japan for the warm welcome extended to members of my delegation, my wife and me since we arrived. I have noted the outstanding support provided by the organizers of this auspicious meeting, in particular the excellent pre-meeting arrangements, which has ensured the smooth passage in light of the distances we have travelled. I wish also to acknowledge the generosity of those Japanese Companies and other International donor agencies for their contributions and the financial support which they have provided for this meeting.

I believe that this is the first Summit ever convened to the address the challenges of Water for the Asia and the Pacific Region. Water or rather the scarcity of it is poses one of the major challenges for several Pacific Island Countries but especially the Smaller Island States (SISs) without river systems and very sensitive ground water resources, typical of atoll island structures. We have always recognized the necessity of conserving our freshwater reserves. The extreme sensitivity, to short periods of drought and seawater infringement, of our freshwater lenses makes it absolutely vital that we do so. The porous nature of our coralline soil also makes the water lenses highly vulnerable to pollutants and contaminants. Therefore water shortages are not unusual and where water is available, maintenance of its quality and suitability for consumption requires additional costs. These problems are further exacerbated by the very high population densities of our urban centres.

It is clear therefore that the impacts of human settlement and climate change on the fragile atoll systems are very significant determining factors on the quality and availability of freshwater sources for our people. Linkages between development, the standard of health and poverty alleviation to water quality and availability can not be ignored. It requires a firm commitment to the preservation and conservation at all community levels and especially by Government. There needs to be strong political will and direction by Government to protect and to use island water resources wisely.

My Government has agreed to use the phrase "Water for healthy Communities, Environment and Sustainable Development" as a theme and this will be incorporated into our national water policy.

This is an attempt on our part to compliment the objectives of this summit. As part of that policy we will also be actively undertaking several initiatives addressing Water and Sanitation concerns. These will be carried out in collaboration with the relevant international agencies (UNDP, UNICEF, etc.)

We appreciate the on-going support of the World Bank in financing the national Kiribati Adaptation Program (KAP). KAP II has been actively engaged at the national level and has notable achievement in several researches and surveys providing valuable data on Freshwater Resources and Water Supply. Other related activities under KAP II will include updating of Water Policy and climate change awareness and implementation of these with particular emphasis on the urban centres (Tarawa Atoll and Kiritimati) and the more vulnerable outer islands (Banaba etc).

In deciding on a such policy it is our expectation that it will compliment and be consistent with regional initiatives and MDGs. At the regional level Kiribati has endorsed the Ministerial Declaration on the 2003 Pacific Regional Action Plan on Sustainable Water Management.

Kiribati also supports and concurs with the United Nations' Millennium Declaration in 2000, "*to halve by the year 2015 the proportion of the world's population who are unable to reach or afford safe drinking water*" *and* "*to stop the unsustainable exploitation of water resources*".

Kiribati is also working towards the target of the World Summit on Sustainable Development in Johannesburg in 2002 of "*halving the proportion of people who do not have access to basic sanitation by 2015*".

Kiribati notes with appreciation the European Community's (EC) endorsement of these goals together with the sustainable management of natural resources, particularly water and energy.

Kiribati is party to the above declarations, and Government reaffirms its commitment to ensure the entire population has access to sustainable safe drinking water and improved wellbeing of the people in general.

Furthermore, my Government is also committed to formulating strategies to design proper sanitation facilities that will limit if not prevent pollution of groundwater and at the same time be community-friendly. We invite donor partners and the relevant international bodies to engage with us in dialogue and technical advice as we prepare to embark on this program.

I am confident that following this summit, we will come away with a better understanding of our individual country positions and needs. I thank the Government of Japan and the participating agencies for having the foresight to initiate this exchange and I invite our development partners to collaborate with us in addressing these lifeline issues.

In closing let me once again express my deepest gratitude for the warm hospitality (in spite of the cold weather) which we have all received since our arrival here in your wonderful country and I also congratulate those involved in the preparations for the excellent arrangements for this meeting.

I wish to conclude by bestowing on all of us our Kiribati traditional words of blessing of Te Mauri, Te Raoi ao Te Tabomoa!

Thank you.

H.E. Lyonpo Dr. Kinzang Dorji
Prime Minister of the Kingdom of Bhutan

His Imperial Highness the Crown Prince of Japan, His Excellency Mr. Yasuo Fukuda, Prime Minister of Japan, His Excellency Mr. Yoshiro Mori, Chairman of the Steering Committee of the 1st Asia-Pacific Water Summit, Excellencies, Distinguished delegates, Ladies and Gentlemen.

At the outset, I have the honour of bringing warm greetings and good wishes from His Majesty the King and the people of Bhutan to this august gathering here in Beppu for the success of the 1st Asia-Pacific Water Summit. May I, on behalf of my delegation, extend our sincere appreciation to His Excellency Mr. Yoshiro Mori, Chairman of the Steering Committee of the 1st Asia-Pacific Water Summit, for graciously inviting us to this historic occasion on water security. We are thankful to the Government of Japan, Japan Water Forum, the Secretariat of the 1st Asia-Pacific Water Summit and the Asia-Pacific Water Forum as well as to the Oita Prefectural Committee for their warm hospitality in making our stay here in the lovely city of Beppu pleasant and most memorable.

We are indeed honoured by the gracious presence of His Imperial Highness the Crown Prince of Japan in our midst today. His Imperial Highness's presence here not only adds testimony to Japan's leadership on water resource management but also demonstrates His Highness's personal commitment to it.

Excellencies, Distinguished Delegates, Ladies and Gentlemen,

In Bhutan, our successive Monarchs have played a dynamic role in steering the kingdom on a path of enormous socio-economic development and subsequently ushering in peace and stability in the country. It was under the visionary guidance of His Majesty the Fourth King Jigme Singye Wangchuck that "Gross National Happiness", was introduced as a guiding principle behind all development programmes, which underscores the concept that development cannot be pursued on the premises of economic growth alone but it should also promote the happiness of the people by supporting their emotional and spiritual well-being. "Gross National Happiness" places individuals at the centre of all development efforts, thereby adopting a multi-dimensional approach in improving the well-being of the people. In giving concrete basis to this development philosophy, preservation and sustainable use of the environment has featured as one of its four pillars, along with economic growth and development, preservation and promotion of cultural heritage and good governance.

Bhutan has received international recognition for its firm commitment on preservation of the environment. His Majesty the Fourth King Jigme Singye Wangchuck was awarded the United Nation's Champion of the Earth Award in 2005, and the J. Paul Getty Conservation Leadership award in 2006.

With an intact state of the environment, Bhutan entered the 21st century, which can be largely attributed to the visionary leadership of our Monarchs who have spurred political commitment towards environmental preservation, as well as promoted our traditional values and way of life that respect nature and all living things.

Bhutan being situated in the fragile mountain ecosystem will be adversely affected due to apparent global warming. Bhutan's economic backbone – hydropower that generates 40% of our national revenue and the livelihood of 71% of the population that depend on subsistence farming will be affected from insufficient water supply, as our glaciers recede rapidly.

As envisaged in the National Environment Strategy of Bhutan, we attach high importance on maintaining the integrity of our watersheds and protecting the existing water resources. The conservation of the wetlands and promotion of integrated watershed development programmes occupy high priority in our development agenda.

As early as 1974, long before the environment started appearing in the agendas of international meetings, Bhutan decided to maintain 60% of the country under forest cover, which was later endorsed by the parliament. More recently it has been enshrined in the draft constitution of the country. Today, Bhutan takes pride in declaring that 72.5% of its land is under forest cover with 35% of which is designated as protected areas. Bhutan's development philosophy that emphasizes on sustainability is strongly complemented by its legislation and policies that ensure sustainable use of resources, promote community involvement in environmental activities, improve land use planning, and most importantly the integration of traditional methods with that of modern natural resource-use practices.

Excellencies, Distinguished Delegates, Ladies and Gentlemen,

The Summit's theme "Water Security- Leadership and Commitment" is particularly relevant and profound in an era where the crisis of water is becoming critical. Although Bhutan is well endowed with rich water resources and the highest per capita per annum water availability of 75,000 m^3 in the region, localized pollution and sporadic seasonal scarcity of water in some parts of the country is a national concern. If definite measures are not undertaken, water could easily become a bone of contention as in some parts of the world. Therefore, with the "water vision" and "water policy" already endorsed by the National Assembly, Bhutan is in the process of formulating a comprehensive "Water Act" for ensuring sustainable management of our water resources.

Striving for the achievement of the Millennium Development Goals, the 10th Five Year Plan of Bhutan that starts next year is primarily focused on poverty reduction. We are proud to acknowledge that we have already achieved the target 10 of the MDG Goal 7 whereby in 2005 84% of Bhutan's population had access to safe drinking water as compared to 45% in 1990. The proportion of people living without access to safe drinking water has declined from 55% in 1990 to 22% in 2000 and 16% in 2003. Such achievement of the MDGs has taken us a step closer to realizing His Majesty the Fourth King's noble aspiration of Gross National Happiness.

The 1st Asia-Pacific Water Summit could not have been convened at a more appropriate time than now to draw attention of the world leaders on the seriousness of water issues in the face of mounting threats of global warming. I would like to commend H.E. Mr. Yoshiro Mori and others who have been instrumental in organizing the Summit. On my part, I would like to assure you the whole hearted support of the Royal Government of Bhutan to the initiatives as well as commitments required for the way forward.

In concluding, I wish to once again express our heartfelt appreciation and gratitude to the Government of Japan and the organizers of this Summit for warmly welcoming my delegation to Beppu. I look forward to a successful Summit with strong commitments from the leaders gathered here today in fostering a common vision in minimizing threats to our collective water security.

Thank You and Tashi Delek.

H.E. Mr. Tommy Remengesau Jr.
President of the Republic of Palau

It is a great honor to attend this 1st Asia-Pacific Water Summit, and I thank Japan, the Asia-Pacific Water Forum and the Steering Committee for the warm welcome and excellent arrangements made for this important event. I particularly wish to give my personal appreciation, and recognition to His Imperial Highness, the Crown Prince of Japan, His Excellency Yasuo Fukuda, Prime Minister of Japan, Mr. Yoshiro Mori, Former Prime Minister of Japan and President of the Asia-Pacific Water Forum, and the many other statesmen, dignitaries and delegates from concerned non-government organizations present for this historic Summit.

There is no more important topic than assuring an adequate and safe water supply to our region. The Summit's priority themes, water financing, water related disaster management and water for development and ecosystems, encompass the urgent water issues and needs of the region well.

Small Pacific island nations, like the Republic of Palau, are fortunate to be able to take part in the planning process and brainstorming. We want to learn and participate, and we want to take an active role in fashioning the solutions for our region, as well as for each of our countries.

Every country, no matter how large or small, depends for its very existence and livelihood on the availability of fresh water. Just as we must carefully manage our marine resources, we must carefully protect and manage our fresh water sources. While it seems to rain a lot in Palau, we simply cannot count on fresh water literally falling from the sky and solving our water management problems.

In Palau, we have several initiatives underway to safeguard our precious fresh water supplies for future generations. I am particularly proud of the Micronesia Challenge, where at least 30% of the near- shore marine areas and 20% of the forest areas of the participating countries and territories, will be effectively conserved and protected by the year 2020. This protection of the forest areas will be of enormous importance to our watersheds.

Another example of Palau's commitment to the environment and to water management is Palau's Water Safety Plan, implemented by the Environmental Quality Protection Board. We also are implementing an Ecosystem-Based Management program, focusing on fostering healthy coastal communities and ecosystems. This project will be implemented in conjunction with the Babeldaob Watershed Alliance. The main goal of this Watershed Alliance is to work with various states in Babeldaob, our largest island, to set aside reserve sites with the focus on clean water management. Sooner, rather than later, we hope to have a watershed protection law in place and to strengthen the protection of current groundwater resources through infrastructure development guidelines.

In addition, we have placed high priority on wastewater management for the improvement of public and environmental health. Through working in this area, we hope to improve public health by reducing diseases and sources of water contamination. We also look forward to reduction of water use through the installation of effective, water efficient, green wastewater systems. Regulations, guidelines and sustained financing are clearly necessary. Unfortunately, Palau currently lacks sufficient technical, human and financial resources in the government and civil society, to fully address these management issues, and we hope to broaden our capacity in this regard as soon as possible.

Palau also is aware of the necessity to prepare for natural disaster emergencies involving water. The large number of drought and tropical storms hitting small islands in recent years has significantly increased the need and demand for services of our National Emergency Management Office, called NEMO. It works closely with the private sector and civil society to ensure that national water rationing would be effectively enforced during times of drought. We need greater capacity in this area, as management of water resources in times of natural disasters must be addressed at the earliest possible time.

Palau will continue to take an active role in our future, and build safeguards and conservation measures into our laws, our future development plans, and into our lifestyles. Summits such as this one will assist us to become more aware of the precise challenges, and to make and implement appropriate initiatives, consistent with our friends and neighboring countries, before it is too late.

This Summit will help bring water issues to the forefront of the regional political agenda. While Palau has achieved a very high percentage of drinkable fresh water most of the time, many of our neighbors are not so fortunate. In fact, there are approximately 635 million people in the Asia-Pacific region who are still without access to safe drinking water.

The region is the most vulnerable in the world with regards to water-related disasters, hindering sustainable development and poverty reduction. Tsunamis, floods, landslides, water-born diseases and epidemics, are on the rise, not diminishing. With climate uncertainties of global proportions, we all must be ready to float in the same very large boat. The sea level is rising, and with this rise, also comes water contamination. We are already suffering the effects of such contamination in our taro patches and wells in our low-lying islands. We look for solutions.

Palau is ready and willing to do its part to safeguard itself, its people, its neighbors, and the Asia-Pacific Community as a whole from water-related disasters. We look forward to a continuing, successful partnership and collaboration with our colleagues here today. Again, I thank H.E. Mr. Yoshiro Mori for taking the lead in this important initiative, and we look forward to continued engagement on these and related issues.

H.E. Mr. Ludwig Scotty
President of the Republic of Nauru

His Excellency Mr. Yoshiro Mori, Chairman of the Steering Committee of the First Asia-Pacific Water Summit and President of the Asia-Pacific Water Forum, Excellencies, Heads of States and Heads of Governments, Distinguished Delegates, Ladies and Gentlemen.

I am extremely pleased to have been invited to participate at this inaugural meeting of the Asia-Pacific Water Summit. I bring with me the warm greetings of the people of the Republic of Nauru on whose behalf I have the honour of addressing this assembly, and the high hopes for a meaningful and decisive meeting.

Water, as we all know, is one of the most basic and essential necessities of our daily life. Hence, it is both alarming and deplorable that 700 million people in the Asia-Pacific region alone continue to have no access to potable water. We, as Leaders of this region, must act quickly and concertedly to remedy this situation.

In the case of Nauru we are more fortunate than others in the sense that we have only a small population of 10,000 citizens, who all currently enjoy access to safe and drinkable water, although in varying degrees of ease and reliability. Having said that, we do however recognize that this situation can easily change and deteriorate in the near future if we do not manage this valuable resource properly and effectively.

Already, water is increasingly becoming scarce in Nauru due to increased and prolonged drought periods brought on by climate change that is affecting the Pacific region more severely. Compounding this problem is the fact that Nauru has no dams or catchment areas for collecting water. As a result, Nauru is totally dependent on water storage tanks for securing water supply.

Potable water on Nauru is provided by two reverse osmosis units. Each of these units is capable of producing around 120 kilolitres of potable water per day. While this is enough water production capacity to meet basic demand, the immediate problem for Nauru is the lack of storage capacity.

Originally, Nauru had a comprehensive water storage system of around thirteen main storage tanks of varying size and structure. Through years of insufficient funding, non-maintenance and exposure to sea air, most of these tanks have become unusable. Today, Nauru's existing storage capacity provides for a maximum of only three days water storage. This is a very insignificant amount when compared to the World Health Organisation's (WHO) recommendation of 20 days of potable water storage per head of population.

This lack of storage capacity poses serious health and safety concerns for the people of Nauru and threatens our long term security and survival. For example, without a secure storage of water, our island is more susceptible to outbreaks of cholera, typhoid and other diseases that will place additional and unaffordable financial and operational pressures on our health budget, which I can readily admit, is already stretched to the limit.

Recognizing the critical state of water on Nauru and the potential implications it can have on our social and economic development, my government has fully committed itself to improving our water management systems. Indeed, water resource management has been identified as a priority and strategic goal in our National Sustainable Development Strategy. Having only been returned recently to office after the August elections, my government will continue to focus our efforts on improving the overall governance and management of water on Nauru.

As mentioned earlier, Nauru's lack of storage capacity is of great concern to the government and people of Nauru. Hence, our recent activities in the area of water development have concentrated largely on improving and securing access to water. Regrettably, any increase in Nauru's main storage capacity is financially prohibitive in the short term due to Nauru's current economic situation and the high cost associated with the purchase and installation of such a facility. However, we hope that with the help of our development partners we will be able to make this a reality someday.

In the meantime, the Government of Nauru has decided to place its immediate attention and efforts on augmenting household water storage capacity. With this objective in mind, Government conducted a survey of household water tanks in the early part of this year, and found that a total of 450 water tanks would be needed to reach a minimum standard of improved household water security for Nauru. In this regard, I am pleased to announce that Nauru, with the kind assistance of Australia, has already purchased 150 galvanized steel water tanks of 18,500 litres capacity. These tanks arrived in Nauru in June and roll-out is almost complete. I am further happy to advise that an additional 45 water tanks of 6,000 litres capacity was kindly donated to the communities of Nauru by the Government of Japan. In making the most of this opportunity, I wish to convey our heartfelt thanks and gratitude to the Governments and people of Australia and Japan for this generous assistance.

The Government of Nauru has also undertaken a variety of projects as part of its wider strategy to improve water resource management on Nauru. These include the purchase of new equipment and engagement of skilled and technical experts in our water delivery infrastructure and water production which has resulted in great improvements in our supply of water to the general populace. With the assistance of regional and international organizations such as the Pacific Islands Applied Geoscience Commission and the Global Environment Facility, we are currently formulating a master sanitation plan which will enable us to administer and protect our groundwater sources and reduce stress on valuable water resources through conservation and wastewater management. We have also undergone studies and assessments to identify alternate water access and reduce our overall vulnerability to rainfall variability.

Although these activities have helped to strengthen and improve the overall governance of water on Nauru, much more needs to be done, particularly in the area of adaptation and mitigation to climate change. Already Nauru, like many other countries in the world, is suffering from unusually frequent and severe bouts of drought. There is overwhelming scientific evidence to support the claim that this is due to climate change.

Much has been said about climate change in recent times hence I will not dwell on the subject for long. But it needs to be made very clear that any water resource management scheme, no matter how carefully planned out and adhered to, will not succeed in ensuring long-term water security if it fails to address the effects of climate change. As noted in the report "Pacific Adaptation to Climate Change – Nauru" done by GEF, UNDP and SPREP, Nauru, as a small island state, and I quote "...has limited resilience to climatic change and therefore the impacts tend to be more severe due to the lack of this natural ability to diversify", end of quote. In Nauru, we have an ongoing project in collaboration with the United Nations Development Programme (UNDP) and the South Pacific Regional Environmental

Programme (SPREP) to identify and implement activities to adapt to climate change and mitigate its harsh effects. But as I mentioned earlier, this is not enough.

Climate change is a global issue and one that poses a major threat to humankind and the continuation of life in our world as we know today. Climate change can only be arrested by the unified and concerted efforts of all nation states. Hence, I call upon all of us here, to renew our commitments and take immediate action to halt this devastating phenomenon. Only then can we truly achieve the long term security and availability of our precious water resource.

Thank you for your attention.

H.E. Mr. Emanuel Mori
President of the Federated States of Micronesia

I am grateful to the organizers of the Summit, most especially the Chairman, the Honorable Yoshiro Mori, for inviting the participation of my country in these proceedings. We are most pleased to be here because the theme of the Summit is fundamental to the livelihood of the people of the Federated States of Micronesia and timely as a development objective of my government.

Indeed, water is intricately linked to the sustenance of life and national development; and for a small island developing state such as the FSM and particularly for the inhabitants of our many low-lying islands and remote and rural areas, water scarcity takes on an urgent call for national action – as well as close collaboration with our development partners. With this urgent need, I have, immediately upon assumption of office six months ago, elevated the priority of the issue of water as a development imperative by assigning its portfolio to the newly created Cabinet-level Office of Environment and Emergency Management.

From an historical perspective, water - the Ocean that is - has been my people's source of strength. The architects of what is now the Federated States of Micronesia recognize the significance of the water as the springboard of our Nation's unity. The water binds us as a people living on disperse islands in the Western Pacific ocean. It does not separate us.

From a cultural point of view, the significance of water is evident in the livelihoods of our villages and communities. It gives meaning to our Micronesian upbringings, whether one aspires to become a farmer, fisherman, or seafarer. As inhabitants of the islands, we have come to appreciate our space and surrounding environment, the make of which are central to our livelihood and essential to our survival.

While we are blessed, with our natural surroundings, we are also faced today with daunting challenges that have now come to characterize how we, as small islands developing states, interact in the global arena. Climate change is the new buzz word around the globe, and it has now taken its rightful place at the forefront of the global agenda. There is no longer doubt in anyone's mind that the adverse impacts of climate change are real and already happening.

Water as a source of life and necessity for economic and social development is impacted by climate change. This is especially true in the island settings where water is scarce. On the average, the streams on the high-volcanic islands are dry for about 20% of a given year, and with global warming looming overhead, they are more likely to be drier in the days ahead.

On those tiny atolls where a third of my people call home, sea-level rise and tidal surges often trigger intrusion of salt water which negatively impacts agricultural crops and ground water sources. This is true as well in the low lying areas of the major islands.

Given the problem of salt water intrusion, many people rely on water catchments, and the minimal amount of water collected is used for drinking, food preparation, and necessary household activities. In extreme events such as droughts, the water collected in a tank for a household averaged 234 gallons per month. That is only sufficient for 3–4 days of supply.

To build resilience to these impacts, we call on our development partners to provide the necessary means and tools. For the small island atolls, these efforts should require transfer of technology in the areas of desalination, renewable energy and other new and innovative technology appropriate for the island settings.

In the year 2000, my country joined the global community in pledging our support to achievement of the Millennium Development Goals (MDGs). One of the targets of the MDGs is to reduce by half the proportion of people without sustainable access to safe drinking water and sanitation by the year 2015. In my view, we are far from achieving that target.

In view of water scarcity, the theme of 'no island left behind' should take on a new significant meaning, one that calls for greater partnership. In this regard, we must time and again reiterate the concept of common but differentiated responsibility. We urge our development partners to rise to the challenge and meet their responsibility under relevant international arrangements.

For our part and in the spirit of the Millennium Development Goals, we are building resilient measures for water scarcity. Our efforts are focused on safeguarding the watersheds by discouraging upland developments and by rehabilitating the coastal lines with conservation measures of mangroves. We have identified protected areas for the purpose of sustaining the environment so that we can conserve the little water resource we have.

At the sub-regional level, the Micronesian jurisdictions have collectively committed ourselves to the Micronesia Challenge where we pledge to effectively conserve 30% of the near-shore marine area and 20% of the forest resources across the Micronesian region by 2020. We appeal to our donor partners for support.

All in all, the issue of water becomes an issue of safeguarding the planet. It becomes a social development issue for those of us who are susceptible to becoming climate change refugees one day. It becomes an economic development issue for those of us whose development are hampered by lack of access to quality water. It is a survival issue for all of us.

In the same spirit that I called for in New York in September of this year, I would like to again appeal to all of us as inhabitants of this Mother Earth, that we are all obliged to be responsible stewards of this creation called Earth. As I said, man has not been doing a good job meeting that responsibility of safeguarding this planet Earth. This and the future generation will be remembered for making a historic turn, in protecting Mother Earth and all that comes with it, including Water as the drops of life.

Kinisou Chapur

H.E. Mr. Mititaiagimene Vivian
Prime Minister of Niue

Your Excellency Mr. Yasuo Fukuda, Prime Minister of Japan, Mr. Yoshiro Mori, Chairman of the Steering Committee of the 1st Asia Pacific Water Summit, distinguished leaders of this fine City, and my esteemed colleagues from the region who are here today, greetings to you all.

Mr. Chairman, let me congratulate you on your being elected as the Chair for this 1st Asia-Pacific Water Summit and to the Secretariat for the excellent arrangements you made for me and my delegation to be able to be here. I would also like to extend through you Mr. Chairman to the leaders and people of this beautiful city of Beppu, my deep appreciation and gratitude for the warm hospitality that we have received since our arrival.

No doubt with your able leadership we will have a successful Summit. We are ready and look forward to actively participate in the forthcoming deliberations on this important agenda that we have before us. I make this statement on behalf of my Government and the people of Niue and in this statement I wish to highlight some important issues on water on Niue and also its importance for the region as a whole.

Mr. Chairman, Water is the essence of life and we in Niue value it greatly as one of the most important resources that we have. We do not have surface running water however we do have our water lens, which is approximately 60 meters underground. Electric pumps are used to extract it for public use. The main concern here for us is the extremely high cost of pumping this water from underground.

Although we currently have adequate water for our needs, our biggest concern for our water is the real threat from projected Sea Level Rise associated with Climate Change. A rise in Sea level will result in saltwater intrusion into the water lens thus reducing the amount of safe water available to the people. This would have grave consequences on all the necessary functions of the nation.

This is a situation that faces all small island countries in the Pacific and unless we do something to mitigate impacts of Climate Change, the future outlook for adequate Safe Water supply for small islands like Niue will be very difficult especially when it is broadcasted worldwide that islands will disappear or become inhabitable because of Climate Change. Therefore we strongly urge those countries that are still yet to ratify the Kyoto Protocol to do so. It is the only means that is currently in place to address Adverse Impacts of Climate Change particularly on the small island states and low lying coastal areas of most developing countries.

Already we have approximately 700 million people in the Asia-Pacific region that have no access to safe drinking water. To add more to that number as a result of the adverse impacts of Climate Change will pose a very difficult challenge for our governments to address now and in the future.

Therefore I see this Summit as a great opportunity for us as leaders from the region to put forward a workable action plan based on the 6 thematic areas being promoted here, namely Water Financing, Water-related Disaster Management, Water for Development and Ecosystems, Developing Knowledge and Lessons, Increasing Local Capacity and Monitoring Investments and Results. I firmly believe that development of good policies for the region based on the above issues would further promote programs to address water concerns for the Asia-Pacific. I would also like to request for the integration of the Pacific Regional Action Plan on Sustainable Water Management plan into the Thematic Areas expressed above as target areas from this Summit.

I am particularly interested in the Thematic Areas regarding Water Development and Increasing Local Capacity for my country. My country is very small and our effort to develop our water resources and increase local capacity is very restricted to basic repairs, water quality monitoring and little training. Already we have 13 productions bore holes in the ground but without pumps. We need the necessary resources to complete this project. This was done as a result of the increase in demand for water from all sectors and also as a response measure to the impact of category 5 Cyclone Heta that nearly destroyed my little county in 2004. In this context my country desperately needs assistance for the Development of its Sustainable Water Supply Program.

Mr. Chairman, I am ready to work with you in the time that we have to have a workable program as a result of this meeting so that we can assure our people in our own countries that we have worked hard and that their welfare with respect to water is well looked after.

I look forward to seeing some solid progress on the work for addressing water concerns for our regions in the very near future as a result of this meeting and I put through a humble request to our developed county partners, business partners, private sectors, and donor partners to assist us.

Mr. Chairman, with limited time available to touch upon every aspect of water and sanitation in small island country like Niue, I would like to encourage everyone with an interest in small island issues to participate in the water and climate session tomorrow. At that session we will have additional technical information available regarding the Water Situation in my country that I would invite you to read through for information at your leisure.

With those few comments I would like to end my speech but before closing I would like to take this opportunity to announce that Niue will be hosting the Pacific Island Leaders Forum in 2008. This Forum will have adaptation to Climate Change as the Theme including the impacts on water resources in small island countries.

I would like to wish the chairman and all the distinguished delegates a very successful Summit.

Thank You Mr. Chairman.

H.E. Mr. Apisai Ielemia
Prime Minister of Tuvalu

Mr. Chairman, Distinguished Guests and Heads of Government.

I take this opportunity on behalf of the Government and the people of Tuvalu to firstly convey our sincere gratitude to the organizers of this very important meeting, for considering Tuvalu's contribution to the 1st Asia-Pacific Water Summit.

I also acknowledge the Government of Japan, for kindly providing funding assistance to my delegation.

Mr. Chairman,

Tuvaluans clearly recognize the importance of its water resources far beyond its domestic and agricultural use. We dearly appreciate its spiritual, aesthetic and recreational values and its fundamental importance to our natural environment and all life. Of all natural resources in Tuvalu like anywhere else, water is regarded as the most important and most precious of all.

When talking about water resources, we often refer to water that we use at home or in other words, we confine our discussions to freshwater, knowing that we also depend on other water sources for our livelihoods.

Freshwater resources in Tuvalu are mainly, rainwater collected from roofs and stored in water reservoirs, groundwater and desalination water. Other water sources are brackish inland water ponds and marine waters. These are the resources that we cherish, but there are factors that impede our efforts in ensuring water security in Tuvalu.

Geographically very small, isolated, remote, scattered and low lying are some of the major factors that contribute to our water security challenge. The development of basic infrastructure is a pressing need and we seek the kind assistance of the international community to provide funding and technical support to help us.

With growing population and increasing development, Tuvalu's water resources are under stress due to the insufficient financial support, water management tools, institutional instruments and infrastructure support to properly conserve, protect and manage our water resources. Again we are seeking partnerships to address these needs.

Global Warming, Climate change and sea level rise, bring about a new set of water related problems to our communities, as it reduces the size of our groundwater sources and simultaneously bring more severe droughts to the country.

Even though, Tuvalu receives an average annual rainfall of approximately 3500 mm, there is a serious issue of water shortage due to the insufficient water storage facilities on all the nine islands of Tuvalu.

Groundwater wells (mainly on the Capital Island Funafuti), is now witnessed to have been abandoned as the water from these wells is getting more salty than before, as a result of sea level rise and contamination from human activities. We need to build long-term storage capacity.

Sanitation on the other hand poses another array of challenges, as we have been impelled to use technologies that were fabricated to suit elsewhere, technologies that would sometimes jeopardizes the security of our water resources. Introduced water sealed latrines and flush septic systems consume a large sum of freshwater, and contribute to the pollution of our groundwater and marine systems.

Water security is repeatedly mentioned in the Government of Tuvalu's Development Plan "Te Kakeega II" as a priority challenge, otherwise my Government has very limited financial resources available to provide enough support to meet the increasing public demand for water. Therefore my Government seeks the opportunity out of this summit for financial and technical support to help us address our water security challenges.

Mr. Chairman,

On behalf of the Government and the people of Tuvalu I would like to convey our deepest gratitude to all donor agencies and governments of different countries that supported our challenge in ensuring water security in Tuvalu. We applaud the quick response from the Government of Japan to our requests in times of drought emergencies. The desalination plants greatly assisted in the much needed water during these droughts. We are also thankful to the recent Australian Adaptation Water funding from which the storage capacity is improved, but more is needed.

On this Note I continue to ask your assistance in providing any assistance possible to improve our capacity to guarantee water security in Tuvalu.

Thank you Mr. Chairman.

H.E. Mr. Dosbol Nur ulu
Vice Prime Minister of the Kyrgyz Republic

Dear Ladies and Gentlemen, Dear participants of the first Asia-Pacific Water Summit,

I would like to express my sincere appreciation to the Japanese government for their warm hospitality and organization of this summit.

The Kyrgyz Republic has considerable water resources: river water, underground water and water accumulated in glacial ice and lakes. The volume of water is about 50 billion cubic meters per year.

In recent years, Kyrgyzstan has been using about eight or nine billion cubic meters per year for personal needs and irrigation, and the rest is used by neighbor countries.

We admit that the technical condition of irrigation and collection drain systems are not satisfactory according to many indicators. A major part of the non-current asset fund (or budget) of the water industry is on the level of physical deterioration.

The Government of the Kyrgyz Republic takes various measures to improve technical conditions and usage of irrigation installations by increasing budget allocation and credit contribution in order to use water resources effectively in the Central Asia region.

Dear Ladies and Gentlemen,

In the Kyrgyz Republic with a territory of 200 thousand square kilometers there are annually on average of 250 extreme natural disasters.

The Kyrgyz Republic with a population of more than five million people suffers about twenty dangerous processes and phenomenon such as flood, breaks of high mountain lake, dams and other similar natural phenomenon. There is also the risk of opening uranium tailing dumps left from Soviet times.

There is real risk that landslides can open tailing dump near the Mayly-Suu River that would have disastrous effects on the entire Central Asian region.

The Government of the Kyrgyz Republic does not have enough potential or financial opportunities for recovery work and preparedness for possible natural disasters. Unfortunately, financial resources for the struggle against natural disasters are not sufficient. Most resources go to salvage operations and not for disaster prevention.

Dear participants of this Summit,

Let us expand on our thoughts on water usage in the Central Asia region and necessary following steps.

The majority of big river flows in Central Asia are formed in territories of the Kyrgyz Republic and Tajikistan.

We have to admit that there are different positions on corporate usage of transboundary rivers.

During Soviet times the large complex of Toktogul reservoir was built. During the process of building 36 hectares of land were flooded for irrigation.

The maintenance and exploitation expenses were covered by the Soviet budget. After obtaining independence Kyrgyzstan is continuing to provide water services to neighbor countries, and incurring losses.

It cannot continue in this way endlessly!

Despite many suggested mechanisms which were prepared in post Soviet times, we have not reached a mutually beneficial agreement for all countries in Central Asia on water resources management.

We continue to seek a mutually acceptable decision on a framework of different international organizations and institutions. Problem solving will require political will of heads of states on an economical approach.

For their part, Kyrgyzstan has initiated the International Water–Power Summit and is establishing the International Water-Power Academy in our capital city, Bishkek.

I would like to point out that this initiative was supported by the European Union. I think the forthcoming Water-Power Summit in the Kyrgyz Republic will be a useful companion to the Asia-Pacific Water Summit.

In conclusion, let me express appreciation to Japan for their warm hospitality and emphasize the significant contribution of the first Asia-Pacific Water Summit in raising world awareness and promoting integration of efforts in problem solving for sustainable and effective management of water resources on the regional and interstate level.

Thank you for your attention!

Concluding Remarks of Speeches by the Heads of Government

Mr. Koïchiro Matsuura
Director-General of UNESCO

Mr. Chairman, Honourable Heads of State and Government, Distinguished Participants, Ladies and Gentlemen.

It is a great honour and pleasure for me to address you on the occasion of the first Asia-Pacific Water Summit. I would like to express my sincere appreciation to the Asia-Pacific Water Forum for convening this important event. I would also like to thank the Japanese Government and the Oita Prefecture for generously acting as host.

This Summit has a very ambitious goal: to elevate water to the top of national and international development agendas.

I fully agree with what has been said today about the formidable water-related challenges confronting Asia and the Pacific. In particular, I would like to echo your concerns about the dramatic impact of climate change on the region. Your remarks underscore the urgent need to act, and to do so together.

UNESCO is fully committed to working with its partners in the region to find workable solutions to water-related issues.

Upon becoming Director-General in 1999, I decided to concentrate the Organization's action on those areas where UNESCO could really make a difference. As part of this new approach, I elevated "freshwater and associated ecosystems" to one of UNESCO's five principal priorities.

At the same time, I advocated that the UN system needed to adopt a much more coherent approach.

More concretely, I proposed the establishment of the World Water Assessment Programme (WWAP), comprising all 24 UN agencies and bodies engaged in freshwater activities.

Before WWAP, UN action in freshwater tended to be fragmented and poorly coordinated. Now, under the Programme, UN agencies are working together to monitor changes in the state of world water resources, and to identify the policies needed to reverse unsustainable practices and negative trends. UNESCO is very proud to host the Secretariat of this UN system-wide endeavour.

UNESCO also helped to set up UN-Water, to further harmonize action across the UN system. UN-Water subsequently adopted WWAP as its flagship programme, and has come to serve as an excellent example of the UN "delivering as One".

WWAP's most important product is the triennial World Water Development Report. The third edition of this Report will focus on the theme of "Climate Change and Water", and is to be released at the 5th World Water Forum in Istanbul in March 2009.

Many scientists from this region are actively involved in the collective process of producing the Report. The new volume is expected to make a major contribution to international scientific knowledge on water and climate change, including the work of the IPCC. It will also set out policy options for water resources adaptation. In all these respects, the Report's findings and recommendations will be of vital interest to the Asia-Pacific region.

WWAP constitutes one of the four main pillars upon which UNESCO's freshwater action is based. Let me turn briefly to the other three pillars, all of which are well anchored in this region, and can make an important contribution to tackling its water challenges.

The first is the International Hydrological Programme (IHP), which has been operating now for over 30 years. IHP is the only science and education programme devoted to freshwater issues in the UN system. Its purpose is to enhance scientific, technical, institutional and human capacities in Member States to achieve internationally agreed water-related goals, including the MDGs.

In the Asia-Pacific region, IHP activities are mainly implemented through UNESCO's Field Offices, and through the IHP National Committees in each of your countries.

Today, this IHP regional network addresses a wide range of water challenges, from the impact of global warming to erosion and reservoir sedimentation issues. Since no one country can solve these problems alone, effective mechanisms for joint regional research, like IHP, are vital.

Indeed, I would like to call on decision-makers to make far greater use of the capacities of IHP National Committees in developing adaptation strategies to cope with the impact of global changes on freshwater resources.

The next pillar of our water-related action is the UNESCO-IHE Institute for Water Education in Delft, The Netherlands. UNESCO-IHE is the largest water education facility in the world, with around 14,000 alumni, nearly 60% coming from the Asia-Pacific region. Almost all of the water specialists trained at the Institute come from the developing world, with the great majority returning to work in their countries of origin. In addition to training students, UNESCO-IHE also collaborates closely with universities and other institutions on various water education, research and capacity-building activities.

UNESCO-IHE stands ready to expand its educational services in order to respond more effectively to the tremendous needs faced by countries in Asia and the Pacific. The Institute has already taken the lead, along with other partners present here today, in developing a Network of Regional Water Knowledge Hubs for knowledge sharing and capacity-building on key water sector issues.

The final pillar of UNESCO's work is the network of seventeen water-related centres that operate under its auspices. These centres, seven of which are located in Asia and the Pacific, are serving major regional and global needs through capacity development and networking.

Let me refer, for example, to the International Centre for Water Hazard and Risk Management (ICHARM) in Tsukuba, Japan, which is working to enhance our understanding of water disasters and to establish effective response strategies.

The Centre helps to identify increased flood risks in different climate change scenarios and provides the Secretariat for the interagency International Flood Initiative. It also assists countries mitigate risks from other water-related disasters through capacity-building and other locally-focused activities such as risk reduction planning.

Notably, ICHARM is the lead organization for this Summit's priority theme of water-related disaster management. Together with the Asian Development Bank, it is currently working on a joint programme proposal devoted to Water and Disasters in the Asia-Pacific region.

Another example is the International Research and Training Centre for Erosion and Sedimentation in Beijing, China. The Centre, which acts as the Secretariat of IHP's International Sediment Initiative, addresses the issue of reservoir sedimentation, a huge problem in Asia.

The other regional centres – such as the Regional Humid Tropics Hydrology and Water Resources Centre for South-East Asia and the Pacific in Kuala Lumpur, Malaysia; the International Centre on Qanats and Historic Hydraulic Structures in Yazd, and the Regional Centre on Urban Water Management in Teheran, both in Iran; the new Arid Zone Water Management Centre in Pakistan; and the HELP Agricultural Water Management Centre in Australia – all contribute to addressing specific regional water-related challenges.

Mr. Chairman,

Let me end with a message from UNESCO. We have always maintained that in order to manage water in a sustainable way, it is necessary to adopt a holistic approach. In the past, technology and the natural sciences have determined approaches to freshwater management. However, it is increasingly apparent that a more integrated strategy is called for – one that recognizes the complex socio-cultural and biophysical factors that influence how water is perceived, used and managed.

This is particularly important in Asia and the Pacific, a region rich in cultural diversity, and with a long history of experience in dealing with water as a fundamental aspect of human life. UNESCO looks forward to working with this Forum to promote a fully interdisciplinary approach to water issues.

A flagship example of our integrated and interdisciplinary approach is the "SWITCH Project" for urban water management in the city of the future, coordinated by UNESCO-IHE. This project, which aims at developing efficient and interactive urban water systems and services for today's new megacities, could provide an ideal response to many of the challenges your countries face.

Therefore, I would like to take the opportunity today to propose a specific "SWITCH-Asia" project, devoted to developing sustainable urban water management strategies for the region's cities.

UNESCO is ready to lead the development and implementation of such a project, in partnership with other stakeholders in the region. We have already received expressions of interest, including for contributions to the around US$50 million required for the project.

Let me add that UNESCO is also providing support to the innovative Asia-Pacific Water Museum project, launched by the National Science Museum of Thailand and the Japan Water Forum. By bringing together engineers, anthropologists and other experts, the Museum will be an ideal window for showcasing this region's valuable traditions of water management. It could also contribute to training and awareness-raising for a "SWITCH-Asia" project.

Mr. Chairman,

If the Asia-Pacific region is to help shape the global water agenda, the outcomes of this Summit must be channelled into the process that leads up to the Fifth World Water Forum in 2009. UNESCO will support you in this endeavour. We are fully committed to the goals of this Summit, and ready and willing to help Member States implement its recommendations through the four water action pillars I have just outlined.

Together, we must and can find the solutions, and effect the changes needed to make the planet a healthier, safer, and more peaceful place for everyone.

Thank you.

CLOSING CEREMONY

Chair's Summary

The Summit was opened by Mr. Yoshiro Mori, Chair of the Steering Committee of the 1st Asia-Pacific Water Summit, which was followed by inspiring remarks from His Imperial Highness the Crown Prince of Japan. Mr. Yasuo Fukuda, Prime Minster of Japan, then addressed the delegates and described how three main water problems were affecting the Asia-Pacific region: the lack of access to water and sanitation, water-related disasters, and the need to secure water for agriculture. He concluded by stating that IWRM is the solution to these problems, and added that water would be on the agenda of next years' G8 Summit.

In his presentation on the Messages from the Asia-Pacific Water Forum, Prof. Tommy Koh, Chair of the Governing Council of the APWF, presented the work of the APWF has been doing since its creation in March 2006 – less than two years ago. Prof. Koh presented the Policy Brief, a report that was prepared by members of the APWF family. He reminded the participants of the urgency of the water-related problems facing the region as well as the innovative work being done in many countries. He urged all Summit participants, leaders and high-level decision-makers in particular, to step forward and commit themselves to making a real difference by bringing water to the top of the political agenda and to work together to solve our region's water problems. This was imperative in order to support efforts at development and poverty reduction.

Sanitation was the focus of the Keynote Speech delivered by His Royal Highness Prince Willem-Alexander of The Netherlands, who also serves as Chairman of United Nations Secretary-General's Advisory Board on Water and Sanitation. Noting that solving water problems in both rural and urban contexts is one of the keys to unlocking economic growth, the Crown Prince went on to make the link between the Priority Themes of the Asia-Pacific Water Forum and the six areas most in need of focus from the Hashimoto Action Plan.

Warm words of encouragement were then given in a video message from United Nations Secretary-General Mr. Ban Ki-moon, who emphasized the critical nature of the region's water problems, especially in the context of climate change, and underlined the urgent nature of this Summit and the need to commit to action.

Finally, in his Memorial Speech, His Imperial Highness the Crown Prince of Japan, offered insightful perspectives about the linkages between water and human development, with historical examples from Japan and from his experiences from his many travels. This enriched the subsequent discourse over the next two days.

Every one of the presentations from the opening ceremony was delivered with a sense of urgency. Coming from leaders at the highest levels, their messages did more than simply set the stage for the Summit. Collectively, they stressed that progress in the region was possible, provided the proper commitment and support was made available.

After the election of the distinguished Chair and Vice Chairs, and the adoption of the agenda, we heard from a number of heads of state and government who talked about water-related challenges in each of their countries and presented some of the solutions they have been working on.

The leaders from the Central Asian countries of Tajikistan and the Kyrgyz Republic described how their countries economies were particularly vulnerable to water-related disasters. They highlighted the need for a cooperative approach to solve regional issues, including increasing water scarcity in the Aral Sea Basin, which is linked to diminishing supplies from glaciers and snowfields and rapid population growth. Increasing investment levels, particularly in rehabilitating their countries aging water infrastructure requires immediate attention. The President of the Republic of Tajikistan and President of the International Fund for saving the Aral Sea introduced a set of new initiatives that would assist in improving water resources management in Central Asia, adding his hopes that the Summit will contribute to their practical implementation.

Noting that water scarcity was an inherent challenge to low-lying islands, mainly because of their lack of river systems and vulnerable groundwater resources, the leaders of the Asia-Pacific small island states of Kiribati, Palau, Nauru, Federated States of Micronesia, Niue and Tuvalu hoped that the Summit would mobilize an increased commitment to financial and technical support. For several of these states, the real threat from projected sea level rise and increased cyclonic activity associated with climate change is a major concern to their lives and livelihoods.

The Prime Minister of the Kingdom of Bhutan described how climate change has led to rapidly receding glaciers, which in turn pose the threat of glacial lake bursts which leads to flash floods that affect human settlements in the downstream valleys.

The Director-General of UNESCO, offered the scientific experience of the organization in seeking solutions to water related problems in the region.

Although clearly focused on water, it is interesting to note that all the speeches by the heads of government identified climate change as a critical challenge to effective and sustainable water management. This clearly demonstrates the need to mainstream adaptation to climate change into the programmes and actions we will be undertaking to solve the water problems of the Asia-Pacific region.

The session on Climate Change, Glaciers, and Water Resources in the Himalayan Region appealed for immediate action to address the problem of glacier melting in the Himalayas, which has been accelerating since the 1970s. Furthermore, insufficient data and information on climate, hydrology, and meteorology for the region, has hindered proper understanding, planning and decision-making. Participants pointed out that a platform will be required for a dialogue on facilitation and financial support with the international community, which would be driven and fully owned by the countries in the region.

The CEO's Roundtable brought together corporate and issue leaders to review the Asia-Pacific Water Scenario, with a focus on the challenges that are having a growing impact on their businesses and their communities, as well as to identify opportunities. The session showcased a number of corporate best practices, offered policy recommendations, and called upon business leaders across the region to support the CEO Water Mandate.

The session on Monitoring of Investments and Results in Water showed how various experiences in the region point to the urgent need to improve policies for the regional and national monitoring of investments and results, and that this monitoring should be linked to priority issues such as investment and its impact on IWRM and water and sanitation.

In the Regional Launch of the International Year of Sanitation 2008, the participants recognized that sanitation, in combination with the capacity to practice hygienic behaviors, is fundamental to the achievement of most of the other Millennium Development Goals. The participants called on Governments to enhance efforts for far more effective implementation of the Hashimoto Action Plan, including the formation of regional fora to address the challenge of sanitation. They also requested that Japan include sanitation on the agenda of the 2008 Toyako G8 Summit.

The Small Islands' Dialogue on Water and Climate called attention to the vulnerability and particular needs of small island countries to effectively manage their water resources under a highly variable and changing climate. One of the main discussion points included the need to change the paradigm for dealing with the vulnerability of small island states from disaster response to disaster risk reduction and disaster management.

In the session dedicated to Ensuring Water Security in the Aral Sea Basin, we learned about the difficulties of trying to balance decreasing supplies with demands from different water use sectors such as irrigation, hydropower, and the environment, particularly in a multi-country and multi-jurisdictional setting. There was however a sense that cooperation was improving and that the rich human resources and technical expertise in the Central Asian region would be a real asset in solving these problems.

The session on Water-related Disaster Management focused the challenges of integrating water-related Disaster Risk Reduction into national development plans. Participants called for the recognition of adaptation to increasing risks from climate change as an issue of the highest priority. The impact of disasters on women and the steps needed to adopt solutions from a gender perspective was an area that required urgent attention.

The session on Water for Development and Ecosystems was devoted to discussing how to secure water for long-term, multi-sectoral needs, generating both development and environmental dividends. The session acknowledged the importance of empowering local actors and fostering good governance as essential means of creating win-win situations in the region. The political dimensions of these developments were recognized as being particularly important. Specific initiatives have and are being launched to take the recommendations forward as mentioned in the Policy Brief 2007.

Recognizing that investing in water is to invest in poverty reduction, the Session on Knowledge, Financing and Capacity Development discussed various finance mechanisms, such as proper cost-sharing between governments and users, and the utilization of the additional resources accrued from the recent rapid economic growth for much needed works for sewerage, sanitation and agriculture.

And, last but not the least, the session devoted to Increasing Capacity for Local Actions presented a series of concrete initiatives, including the creation of capacity building hubs for increasing local capacities. These hubs will serve as an active link with the various country's NGOs, CSOs and local governments to support advocacy towards the three APWF Priority Themes.

The ten Sessions were much more than a mere forum for discussion. They were critical stepping stones towards achieving progress under the three Priority Themes and five Key Result Areas of the Asia-Pacific Water Forum. A number of Concrete Initiatives were announced, and these will ensure that progress will be made towards solving our region's water problems. There is a clear sense of a real political commitment to change, with innovative programmes and initiatives in many parts of the region. However, it was also made clear during several of the sessions that strengthened support from the region's leaders and decision makers is absolutely essential in order to attain the levels of progress needed to improve the lives and livelihoods of hundreds of millions of people in our region.

The Summit's closing ceremony began with reports from the session organizers. The President of the World Water Council and a representative of the government of Turkey, invited participants to the 5th World Water Forum in Istanbul (March 2009). Delegates from the Americas and Europe addressed the opportunities for partnership with the Asia-Pacific region. They were followed by the Chief Executive of PUB Singapore who acknowledged and thanked the Chair of the 1st APWS for Japan's kind invitation for Singapore to host the 2nd Asia-Pacific Water Summit.

At the end of the ceremony, Mr. Ravi Narayanan, APWS co-rapporteur, read the Chair's Summary. Ms. Erna Witorlar, the other Summit co-rapporteur, read the "Message from Beppu", which was unanimously accepted by the participants. Mr. Tetsuzo Fuyushiba, Minister of Land, Infrastructure, Transport and Tourism, then delivered a speech on behalf of the Japanese government, stating how measures for adapting to climate change were important.

The closing ceremony ended with a closing address by Mr. Yoshiro Mori, Chair of the 1st APWS, who expressed his profound appreciation to all those who had committed themselves to the 1st APWS, and expressed his expectations for participants' future activities.

The 1st Asia-Pacific Water Summit, held 3–4 December 2007, in Beppu, Japan, was attended by ten Heads of State and Government, 32 Ministers, and representatives from over 36 Asia-Pacific countries. In addition to Mr. Mori (Chair), H.E. Mr. Emomali Rahmon, President of the Republic of Tajikistan, H.E. Lyonpo Dr. Kinzang Dorji, Prime Minister of the Kingdom of Bhutan, H.E. Mr. Anote

Tong, President of the Republic of Kiribati, Hon. Mr. Lim Kean Hor, Minister of Water Resources of the Kingdom of Cambodia and Meteorology and H.E. Mr. Hu Siyi, Vice-Minister, Ministry of Water Resources of the People's Republic of China served as Vice Chairs of the Summit. The Summit's opening ceremony was attended by 880 people, including 140 invited Summit participants (72 from abroad) and attracted 260 members of the press (40 from abroad).

Closing Remarks

H.E. Mr. Tetsuzo Fuyushiba
Minister of Land, Infrastructure, Transport and Tourism

It gives me great pleasure to witness the success of the First Asia-Pacific Water Summit, mirrored by the attendance of so many of you from all the different countries of the region. It is now time to move on to the closing ceremony. On behalf of the Government of Japan, I would like to express our sincere gratitude to all those of you who have committed their time and efforts in making this conference a success.

This summit is historic because it marks the world's first heads of government conference that deals with water as its main theme. We are also glad to see that so many people attended the open-events that were held between December 1 and December 5. It is an honor for us to have so many active participants.

I strongly hope that the Asia-Pacific region will play an engine role for the world's economic growth and that the sharing of common water issues will strengthen the regional coordination and help us to develop together.

And now we will bring the First Asia-Pacific Water Summit, which ran for two days, starting on December 3, to an end with the closing ceremony. I would like to express my deep gratitude and respect to everyone who has taken an active part in the Water Summit right up to the end, to the members of the Summit's Steering Committee, especially to Former Prime Minister Yoshiro Mori, and to everyone's endeavor.

Over the past two days, you will have had the opportunity to soak yourselves in Beppu's hot springs and I hope you have enjoyed Japan's traditional resort atmosphere here. At a time when the world is entering an era of exchange, the Government of Japan is aiming at making Japan a country for tourists. And I am the Minister in charge of the task.

I sincerely hope that everyone who is present here will have had opportunities to deepen their understanding of Japan and its culture and will want to make a longer visit to this country in the future. Thank you very much.

SESSIONS

Climate Change, Glaciers, and Water Resources in the Himalayan Region

organized by International Centre for Integrated Mountain Development (ICIMOD)

Objective

The river basins of the nine rivers originating in the greater Himalayan region – also known as the 'Third Pole' – sustain more than 1.4 billion people and meet their needs for water for irrigation, hydropower, household and other uses. Climate change is expected to have serious consequences on water-related eco-system services provided by these resources and will have impacts on livelihoods in this region. Climate change may increase the frequency and magnitude of glacier lake outburst floods (GLOFs) and flood-related disasters. The levels of water stress may also increase due to seasonal and annual variability in water availability at different locations in the region.

The glaciers of the Himalayas, especially in the eastern and central regions, have been shrinking at an accelerated rate in recent decades, compared to the 1970s. A retreat rate of about 10 to 30 metres per year is quite common, and, in some cases, retreat rates are much higher. The retreat rate for the Imja glacier near Mt. Everest has increased from 41 metres per year from 1962–2001, to 74 metres per year from 2001–2006. The rate of retreat of the Gangotri glacier in the Indian Himalayas over the last three decades was more than three times the retreat rates during the preceding 200 years. On the Tibetan Plateau, the glacial area has decreased by 4.5% over the last 20 years, and by 7% over the last 40 years.

Such drastic reduction in ice cover has not, however, been observed in the northwestern Himalayas, Karakorum, Hindu-Kush, or Pamir. In the '70s, many of the glaciers in these mountain ranges have exhibited short-term thickening and expansion. Throughout much of the '80s and most of the '90s glacier retreat and thinning became the trend, just like in the period from 1920 to the 1960s.

These observations have been largely based on limited case studies, however. There is a need to reduce scientific uncertainty by assessing and monitoring environmental changes in the Himalayan region. As most of these glaciers and the impacts of their retreat transcend political and geographic boundaries, monitoring changes will be more effective through regional cooperation, primarily in: (a) sharing information for assessing and monitoring environmental changes, and (b) defining common interests among countries of the region for managing the risks of glacier- and water-related disasters.

The objective of this session is to discuss issues related to the consequences of climate change (a) for glacier- and water- related disasters, and (b) for water stress due to seasonal and annual variability in water availability in the region.

Agenda

The session on "Climate Change, Glaciers, and Water Resources in the Himalayan Region," was graced by the presence of His Imperial Highness the Crown Prince of Japan and that of His Excellency Yoshiro Mori, the former Prime Minister of Japan.

CHAIR: Professor Nobuo MIMURA, Ibaraki University

Welcome Remarks by Dr. Andreas SCHILD, Director General, ICIMOD

Keynote Address by Prof. Said-ud-Din SOZ, Minister of Water Resources, India

Keynote Address by H.E. Mr. HU Siyi, Vice Minister of Water Resources, People's Republic of China

Keynote Address by H.E. Mr. Junzo YAMAMOTO, Vice Minister, Ministry of Land, Infrastructure and Transportation, Japan

Keynote Address by Mr. Ken NOGUCHI, Member, Steering Committee of the 1st APWS

Introduction of the Panelists and Introductory Remarks by the Moderator, Dr Andreas SCHILD, Director General, ICIMOD

PANELISTS:

- H.E. Mr. Lyonpo Dr. Kinzang DORJI, Prime Minister of the Kingdom of Bhutan

- H.E. Mr. Gyanendra Bahadur KARKI, Minister of State for Water Resources, Nepal
- Dr. Jamal SAGHBIR, Director, The World Bank
- Dr. Kuniyoshi TAKEUCHI, Director, ICHARM

Remarks by Mr. Ariyuki MATSUMOTO, Vice President, JICA

Closing Remarks by Professor Nobuo MIMURA, Ibaraki University

Summary of Discussions

Context

Mountains are the earth's unique freshwater reservoirs. They store immense amounts of water and hold them to gradually release to support the lives downstream. The important cooling and mitigating climatic services offered by the frozen ice of the mountains to the atmosphere have been taken as a given fact of nature. Despite their importance for the well-being of humanity, mountains receive little attention in the international development agenda.

Climate change has brought about other dimensions to the awareness of the scientist, and a growing number of citizens and politicians. The vision of melting glaciers and of mountains falling down because of receding permafrost has become a frightening vision. It questions the vision of 'mountains built for eternity' and the growing concern is for the sustainability of systems we have all taken for granted. This refers to the familiar panorama of snowcapped mountains, not only as a source of beauty but also as a heritage for mankind. It refers also, and very realistically, to the availability of water, and electricity, and touches the basic requirements and services we need for our survival. All of a sudden, mountains as a natural system with natural and human features are not anymore in a position to provide the services we have taken for granted.

The greater Himalayan region known as the Water Tower of Asia, directly sustains more than 150 million people, but the river basins of the nine rivers originating in the region count 1.4 billion inhabitants, and up to 3 billion people live from the food and energy produced by the Himalayan rivers. The Himalayas are also considered as a 'third pole', as here lies the greatest concentration of snow and ice outside of the pole. However, the storehouse of fresh water has been degrading over the years due to global warming and climate change.

Content

The session was addressed by H.E. the Prime Minister of Bhutan and the ministers of China, India, Japan and Nepal, along with the representatives of JICA, the World Bank, ICHARM of Japan, and the Steering Committee of the 1st Asia-Pacific Water Summit, as keynote speakers and panelists. The details are summarized below in the form of a matrix.

Country/ Organization	Scientific Uncertainty	Adaptation	Initiatives	Regional Dialog	International Community Engagement
Selected Regional Member Countries of ICIMOD					
Bhutan	Concern over GLOFs; especially because of hydropower generation, Bhutan's economic backbone.	Comprehensive framework for adaptation and mitigation being pursued.	Successful completion of the National Adaptation Program of Action (NAPA).	Commitment to cooperate at the regional level for information-sharing and for risk-management.	
China	Concern about the impact of climate change on China's water resources.	Pursuing adaptive strategies to tackle natural disasters and realize sustainable utilization of water resources.	Integrated river basin management; development and promotion of technologies for water-saving and sea-water utilization; and environment-friendly hydropower development.		
India	Concern over the lack of authoritative data.	Pursuing the development of programs for adaptation to climate change.	Ministry of Water Resources' National Institute of Hydrology studying glacier melting with support from scientists at the world-class IITs of India.		
Nepal	Serious concern over GLOFs in Nepal and in China (Tibet) on shared river basins; hydropower development vital for Nepal.	Pursuing the development of programs for adaptation measures; Himalayan Climate Change Study Centre envisaged in Nepal's Water Resources Strategy.	Early warning systems and GLOF-mitigation measures established at the Tso Rolpa glacier lake to protect communities and infrastructure, mainly hydropower plants.	Commitment to regional cooperation.	

Country/ Organization	Scientific Uncertainty	Adaptation	Initiatives	Regional Dialog	International Community Engagement
Himalayan Region's Major Bilateral Donor Country					
Japan	Concern over the recent Bangladesh event and the role of climate change, if any.	Adaptation measures necessary with A 'zero victim' goal.	Japanese disaster risk reduction experts in Nepal as early as 1977; established Sabo Center in Nepal in 1991.		Japan interested in contributing further for technical assistance.
Himalayan Region's Major Bilateral and Multilateral Donor Agencies					
JICA	Urgent need from donors to properly assess the extent of the adverse effects brought about by recent climate change.	JICA to support adaptation measures to mitigate anticipated negative impacts due to climate change.	JICA academics and geological engineers involved in Nepal from 1990 to 1996 and in Bhutan from 1997 to 2005 to assist with glacial lake problems.	Role played by ICIMOD as a regional international organization to be highly instrumental for coordinating related countries.	JICA currently discussing initiatives for possible adaptation measures; expected to be discussed by Japan at the G-8 Hokkaido Summit.
World Bank	Very little scientific cooperation and data-sharing on Himalayan research; cooperation inevitable, benefits of cooperation higher than those achieved through unilateral pursuits.	Need to pursue adaptation as a Human Security issue.	World Bank ready to support regional cooperation for establishing regionally-shared databases and for disaster preparedness and risk management.	Platform for a regional dialog driven and fully owned by the countries in the region urgently needed to understand the benefits of cooperation and the costs of non-cooperation.	Long-term commitment of international community for technical and financial support necessary.
Selected Stakeholders in Japan					
APWS (Alpinist Noguchi)	Personally observed significant changes first in 1992; strange signs observed since 1998.		GLOF mitigation measures essential. Need concrete action urgently.		
ICHARM	Need to establish observation networks in the Himalayas.	Japan's Sabo experience helpful for adaptation measures; some similarities with GLOFs.	Major activities at ICHARM on early warning systems; also, a good program on hazard mapping.		

Country/ Organization	Scientific Uncertainty	Adaptation	Initiatives	Regional Dialog	International Community Engagement
ICIMOD's Current and Future Role on Climate Change, Glaciers and Water Resources in the Region					
ICIMOD	Can play a role to enhance scientific understanding through glacier monitoring, flood forecasting, regional database creation, and impact assessment.	As a center dedicated to mountain development, can help capacity building in Regional Member Countries to develop adaptation measures in a mountain context.	Have developed inventory of glaciers and glacier lakes in the Himalayan region; have been monitoring selected glacier lakes using satellite-based remote-sensing techniques; recently started field-based monitoring in collaboration with Keio University of Japan.	Can help facilitate regional dialog.	Can help facilitate discussions in international forums.

Messages

1. Unless the international donor community has a correct appreciation of the magnitude and scope of the problems related to climate change, glacier melting, and water resources in the Himalayan region, and take action urgently, the consequences of climate change for glacier- and water- related disasters and for water stress due to spatial and temporal variability in water availability, may bring catastrophic results to the livelihoods of more than 1.4 billion people living in the nine river basins of the region.
2. A platform will be required for a dialog on facilitation and financial support with international community, which will be driven and fully owned by the countries in the region. In this process, ownership and commitment by the national governments will be critical to the success of the dialog. Long-term commitment of the international community will also be necessary for the success of this process, which by its very nature tend to be long and arduous.
3. There is little data and information on climate, hydrology, and meteorology for the Himalayan region, and this has hindered proper understanding, planning and decision-making. Because of this dearth, the Fourth Assessment Report of IPCC has categorized the region as a 'white spot' on the global climate map. Given the global ecological significance of the region and the importance of scientific data and information, it is essential to create regional databases on different aspects of mountain regions as well as strengthen regional cooperation for the exchange of data and knowledge (a) to reduce scientific uncertainty about the region and (b) to facilitate proper planning and decision-making by the national governments

in the region and by international organizations. Specifically, the National Adaptation Plans of Action (NAPAs), currently being prepared at the initiative of the UNFCCC, need to suggest plans of action that could help improve data availability in the region.

Concrete Initiative

Initiative Title: Climate Change, Glaciers, and Water Resources in the Himalayan Region

Specific Objectives

1. Development of monitoring and assessment schemes for cryosphere and water resources in a regional scale; and strengthening national capacities for the monitoring and assessment of cryosphere and water resources on a national and field-based scale.
2. Capacity development among ICIMOD's regional member country (RMC) partners to deal with multi-hazard risk assessment, including vulnerability assessment at community levels and the development and application of early warning systems.
3. Development of regional information sharing mechanisms, including the establishment of a regional water database and the development of basin-wide scenarios on climate change and its consequences for water availability.

Implementing Organization

International Centre for Integrated Mountain Development (ICIMOD)

Active Partners at Country Level

Regional Member Countries of ICIMOD (Afghanistan, Bangladesh, Bhutan, China, India, Myanmar, Nepal and Pakistan)

Implementing Schedule Before and After the 1st APWS

Before the 1st APWS: proposal development

After the 1st APWS: Planning, design, fund mobilization, and implementation

At the 5th World Water Forum in March 2009: Presentation of the progress report on the concrete initiative.

How to Assess the Performance/Progress

Monitoring of outcome, indicators, actions, and sub-actions

How will the Initiative be Funded and Sustained

ICIMOD has a long-term view of a program based on the cryosphere-level water resources as one of the critical elements in its thrust area on Integrated Water and Hazards Management (IWHM). Therefore, ICIMOD can plan and implement this initiative with funds mobilized from donors and can then sustain through its own ongoing programs.

CEOs Mandate for Action in Water

organized by United Nations Global Compact

Objectives

- To communicate CEOs' key messages on the role of Asia-Pacific's private sector in meeting the water challenge before a select audience of policy makers, civil society representatives and business colleagues
- To showcase their companies' best business practices
- To offer policy recommendations from the private sector to attending policy-makers
- To call on other business leaders in the region to explore participation and commitment to the CEO Water Mandate

Agenda

Welcome Remarks by the Moderator, Ms. Charmine Koda, Director of the United Nations Information Centre

Introduction by Mr. Fred Dubee, The United Nations Global Compact, CEO Water Mandate

Presentation on "AP Water Scenarios: Challenges and Opportunities" by Ms. Catley-Carlson, Chair of GWP

Round Table Discussion

PANELISTS:

- H.E. Mr. Emanuel Mori, President of the Federated States of Micronesia
- Mr. Khoo Teng Chye, Chief Executive, PUB Singapore
- Mr. Shigemi Tamura, Chairman, The Tokyo Electric Power Co., Inc.
- Ms. Julia Marton-Lefèvre, Director General, IUCN
- Mr. Michio Takarada, Chief Executive Officer, Grundfos Pumps K.K.
- Ms. Sherisa Pulido Nuesa, Chief Finance Officer, Manila Water Company
- Professor Tetsuro Tsujimoto, Nagoya University
- Dr. Annie Wu, Chair, World Trade Centers Association Greater China
- Mr. Pierre Victoria, Director International Institutional Relations, Veolia Water

- Dr. Yumio Ishii, Chair, CTI Engineering Co., Ltd
- Dr. James Horne, Deputy Secretary, Water Group, Australian Department of the Environment and Water Resources
- Dr. Masaru Kurihara, Advisor, Toray Industries, Inc.
- Dr. Don Blackmore, Chair, eWater Cooperative Research Centre

Summary of Discussions

Under the aegis of the United Nations Global Compact, the APWS CEO Panel/ Round Table provided a platform for 16 CEOs and other leaders from the Region and international community to discuss the Regional Water Scenario with focus on challenges and opportunities and to communicate clearly their commitment to address critical water issues at the strategic, policy and operational levels. Case studies including the creation of corporate supported networks to help improve the effectiveness of small water operators in poor regions, the development and dissemination of membrane technology to improve the cost effective delivery of quality water as well as reducing desalination energy requirements and the use of system loss reduction savings to significantly and sustainably increase potable water access were presented. The discussion outlined the tendency toward increased emphasis on "Public Private People" partnerships as well as a greater and more profound commitment to multi-stakeholder involvement in tackling challenges.

The concrete proposals tabled at the Round Table included: the implementation of a water knowledge bank in Singapore, a multi-stakeholder training and development resource in China, a comprehensive study on "Asia Pacific Water and Business Challenges and Best Practices" and exploration of increased engagement with the United Nations Global Compact CEO Water Mandate by business in the Region.

Ms. Margaret Catley-Carlson: Asia-Pacific Water Scenarios: Challenges & Opportunities

Business world is waking up to the reality of water. The three Ps – population, prosperity and pollution – are now making the situation very difficult for water. So we are all waking up to the reality of water. Technology & science is just a part of the solution. Most important role for business, a part of finding solutions in how to be using water better. Look at the water footprint to see how it can be improved, and how each part of the business might be contributing to the solution. Essential for business to look at their water use patterns and take part in dialogue about the adequacy of water in their practices. Don't reinvent the wheel, use existing materials. Be a part of your own water basin.

Mr. Fred Dubee: CEO Water Mandate

How do we harness the creativity of business for water issues? Background for CEO Water Mandate.

6 areas of focus

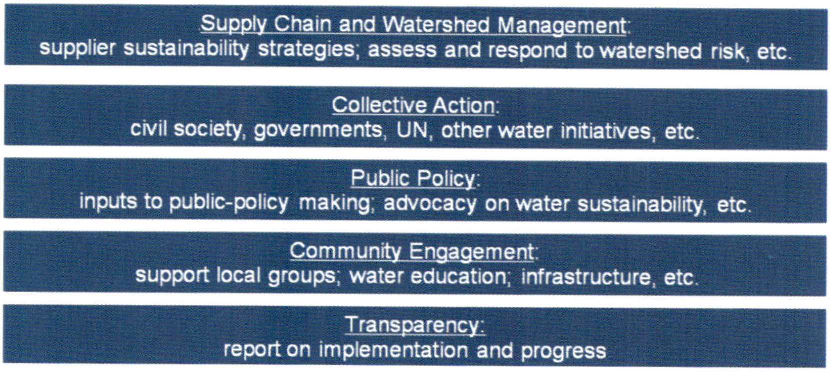

Mr. Khoo Teng Chye (CEO, PUB Singapore): Private sector can address water challenges at 2 levels: the local level and the international level. Local level: In Singapore, industry consumes a great amount of water (almost half), several programmes targeting the industrial sector (large scale recycling of water for industrial purposes (New Water), water efficient buildings, close collaboration with private sector on technology and R&D, community engagement programmes, water network to advise on policy, etc.). International level: PUB is a member of the Asia-Pacific Water Forum: knowledge- hub as a focal point for technology learning and sharing, Singapore International Water Week, a global platform for stakeholders to convene in Asia. PUB endorses and supports the CEO Water Mandate.

Mr. Tamura, Chairman, Tokyo Electric Power Company (TEPCO): Hydroelectric power plant (10% of electricity generated) introduces TEPCO's contribution to water challenges (natural reserve, water allocation, through a partnership with government TEPCO's hydro-dams are also available to help control flooding and drought events, etc. Water infrastructure (dams, heat pumps) needs to be made multi-purpose to serve society at large.

Ms. Marton-Lefèvre, IUCN: New programme on business & biodiversity. IUCN is advising multinationals to integrate ecosystem thinking in their business practices and is growing their partnership efforts to increasingly work with the private sector on issues of biodiversity. In Mekong River implementing a dialogue programme to involve all relevant stakeholders to secure the sustainability and integrity of water.

Mr. Takarada, Grundfos: 20% of power consumption is used to generate pumps in the world. New models need to be at least 5% more efficient, pollution reduction efforts in the pump manufacturing process, solar powered pumps have been developed, etc. Consumers still prefer cheaper goods to environmentally-friendly goods, and awareness programmes are needed to educate consumers.

Ms. Nuesa, Manila Water: This company is a result of a privatization initiative in 1997 for water supply and wastewater management. The company is addressing issues in direct operations, having recovered some 500 to 600 million liters of water losses since the privatization, partnering with government and NGOs in watershed protection and developing new water sources, engaged in joint projects with the World Bank, IFC, ADB and the Japan Embassy, raising funds for "water for the poor" projects, and having published its 1st Phil. Sustainability Report (GRI).

Prof. Tsujimoto, Nagoya University: speaking on behalf of Mr. Kawaguchi (Chairman, Chubu Economic Federation). A recently held Symposium in a leadup to the 1st APWS which adopted a declaration focusing on the need for multi-stakeholder collaboration efforts on water issues and that industry in particular ought to be involved from the planning stages of society's projects.

Dr. Wu, Hong Kong, China: Pearl River Water Center to study water issues, especially for domestic water supply. Support the CEO Water Mandate. Dr. Wu issues a call to leaders to strive to work in public/private partnership for the resolution of water issues so that public policy serves people better.

Mr. Victoria, Veolia Water: Our main job is to manage water and to supply it while saving water and money. Our main challenge is to improve the trust between all stakeholders, as we need to collaborate to improve water systems: how to share our technology, find new adapted technologies, and to transfer our company's know-how to NGOs and other operators. Veolia is involved in Cambodia in a programme with UNICEF and other partners called the Mini-Networks on Drinking Water. The four main challenges we need to work on together are: efficiency, transparency, integrity and capability.

Dr. Ishii, CTI Engineering: Sumida River revitalization project, which required increasing taxes, that Tokyo citizens agreed to pay. This is how Japanese people managed to have clean water. Demands on water are increasing so the need for this type model whereby citizens will have to pay for cleanup will be increasing too. Since clean water means high prices of commodity, businesses must therefore demand citizens to pay high prices for commodities to keep the water and air clean.

Dr. James, Department of Water, Australia: Role of public/private partnerships. In light of climate change impacts, Australia currently has a number of water infrastructure projects planned (desalination and recycling plants, dams, etc.). While there is a role for public/private partnerships in Australia, the government still has an important role to play (regulation, planning, providing incentives, etc.).

Dr. Kurihara, Toray, Japan: Membrane technologies' contribution to water issues, which enable control of high precise water quality and high speed treatment. Thanks to membrane technology, Japanese companies manage to recover 80% of the water used in their processes, hence reducing discharges. Membrane technologies are: capable of appropriately treating quality and quantity of water from various water resources, from small-scale decentralized systems to large-scale centralized ones and which are already in use throughout the world.

Dr. Blackmore, eWater Cooperative, Australia: One of the things that communities want is reliable information that can be trusted and eWater is building water management tools to achieve this. In the mid-90s in Australia, the irrigation industry was going broke, and the government took the bold step of liberalizing the industry, which has been successful, as demonstrated by the wine exports which have grown from $200 million in 1993 to $2 billion today (that could not have grown such without liberalizing the irrigation industry).

Messages

- Harness the fortune at the bottom of the pyramid: The companies represented at the CEO panel have given great examples of how to be aware and creative in harnessing the "fortune at the bottom of the pyramid", which in most instances is currently being lost.
- Involve people: The theme of public - private – people (PPP) was central to the discussions of the panel, whereby, people are the stakeholders and the involvement of people is fundamental.
- Build trust: Another key point raised by the panel is the crucial need to build trust when involving stakeholders.

- Harness the creativity of business: The examples brought to the CEO panel show how it is possible to harness the creativity of business to make technology productive, whereby business can indeed take extraordinary technology and make it serve the needs of people.
- Share knowledge better: The examples of the knowledge hub in Singapore, the intent to create a knowledge water center in China, the water research center in Australia, are examples of how this can be achieved.
- Stakeholder dialogue that empowers: through educating consumers, educating industry, and educating government.

Concrete Initiatives

Initiative Title

- World Economic Forum
- Global Compact Conference for CEO Water Mandate Signatories, United Nations, New York, March 5, 2008
- Pearl River Delta Conference, Nansha, China Arpil/May, 2008
- Singapore International Water Week, Singapore, June 2008
- Hunan
- Study on business & water in the Asia-Pacific Region

Specific Objectives

To call companies to action, through support of the CEO Water Mandate.

Implementing Organizations

- Individual companies and organizations
- Support from Global Compact Office and Local Networks

Active/Potential Partners at Sub-Regional Level/Country/Sub-Country

- Global Compact Local Networks
- National Business organizations
- Key NGOs: IUCN, Global Water Partnership, WBCSD

Monitoring of Investments and Results in Water: Key Result Area 4

organized by United Nations Economic and Social Commission for Asia and the Pacific (UNESCAP)

Objective

- To solicit views and comments from senior officials from countries and international organizations in the region attending the First APWS in order to improve the recommendations of KRA4;
- To mobilize support from the above-mentioned senior officials with a view to ensure consensus on the recommendations of KRA4; and
- To seek advice and support for initiatives resulted from the series of studies of KRA4.

Agenda

Welcome Statement by Mr. Rae Kwon Chung, Director of Environment and Sustainable Development Division of UNESCAP

Opening Statement by H.E. Mr. Jose T. Atienza, Secretary of Department of Environment and Natural Resources of the Philippines

Introduction of the findings and recommendations of KRA4

Summary of relevant results of the 3rd Southeast Asia Water Forum (3rd SEAWF) by Dr. Lee Jin, Vice Chair of Malaysian Water Partnership and Programme Coordinator of 3rd SEAWF Organizing Committee

Keynote speech by Dr. He Changchui, Assistant Director General of FAO on behalf of the UN Water

Comment by the Special Guest, H.E. Dr. Ram Sharan Mahat, Finance Minister of Nepal

Panel Discussion on the recommendations

PANELISTS:

- Mr. Thierry Facon, FAO

- Dr. Lee Jin, 3rd SEAWF
- Dr. Woo Chong Um, ADB
- Mr. Kazushi Hashimoto, Executive Director, JBIC
- Dr. Siripong Hungsaprug, Thailand
- Mr. Terrence Thompson, WHO

MODELATOR: Mr. Rae Kwon Chung, UNESCAP

Presentation of KRA4 initiatives by UNESCAP

Panel Discussion on KRA4 initiatives

PANELISTS:

- Mr. Woo Chong Um, ADB
- Mr. Thierry Facon, FAO
- Mr. Kazushi Hashimoto, Executive Director, JBIC
- Dr. Siripong Hungsaprug, Thailand
- Mr. Terrence Thompson, WHO
- Mr. Quamrul Islam Siddique, President, Bangladesh Water Partnership

MODELATOR: Mr. Rae Kwon Chung, UNESCAP

Presentation of the Session Report by UNESCAP

Plenary discussion moderated by Mr. Rae Kwon Chung, UNESCAP

Summary of Discussions

The Session on Key Result Area 4 on Monitoring of Investment and Results was well attended. It benefited from the participation of H.E. J.T. Atienza, Secretary of Department of Environment and Natural Resources, who initiated the Session with his inspiring opening statement on the importance of monitoring of investment and results in the context of socio-economic development of the Philippines and also of ASEAN. The Session also benefited from the participation of H.E. Dr. Ram Sharan Mahat, Finance Minister of Nepal as a special guest, who provided his perspectives on this subject in the context of development of his country and expressed support to the recommendations of the KRA4 study.

The Session built on the earlier efforts of the Regional Workshop on Monitoring of Investment and Results at the Third South-East Asian Water Forum (SEAWF-3), which led to a set of recommendations adopted by the South-East Asian Ministerial

Meeting on Water Resources on 27 October 2007 and presented by Dr. Lee Jin, Vice Chairman of the Malaysian Water Partnership in his capacity as the Programme Coordinator of SEAWF-3 Organizing Committee. Apart from the recommendations adopted at the SEAWF-3, the experiences of UN Water were also presented by the Assistant Director General of the Food and Agriculture Organization of the United Nations. The first panel discussion focused on strategies to enhance regional capacity on monitoring, in which most of the discussion pointed to the need to ensure not only the productivity of water utilization, as suggested by H.E. Mr. Ohn Myint, Deputy Minister of Agriculture of the Union of Myanmar, but also the integrity of ecosystems. Following the presentation of the initiatives identified at the SEAWF-3, the second panel discussed ways forward, in which almost all panelists and participating experts agreed with the main ideas of the initiatives and suggested ideas to improve their feasibility.

The detailed recommendations are summarized as follows:

a. Water Supply and Sanitation
   1. Establish a Regional Water Supply and Sanitation (WSS) Information System
   2. Establish programmes to strengthen monitoring of water supply and sanitation sector at the national and sub-national levels in the Region
   3. Establish an APWF WSS Monitoring Implementation Task Force

b. Agricultural Water Management
   1. Need to improve irrigation services and maintain infrastructure investment
   2. Need to improve efficiency of existing government-owned programme
   3. Need to adjust investment strategies to meet food securities through monitoring
   4. Need to harmonize irrigation approaches and develop a set of common indicators for governments and donors

c. Ecosystem Management
   1. Need to establish monitoring system on the status of rivers and ecosystems
   2. Need to mobilize local authorities to develop plans for rehabilitation and restoration of rivers and ecosystems
   3. Need for common guidelines to harmonize local and national efforts in rehabilitation and restoration of rivers and ecosystems

d. Water-related Disasters Management
   1. Need for standardized methodologies to assess socio-economic impacts of water-related disasters
   2. Need for increased investments in measures to address water-related disasters

e. Integrated Water Resources Management
   1. Need for National-level mechanisms for monitoring investment and results for the entire water sector
   2. Establish a Council of Ministers on Water of the Asia-Pacific (CoMWAP) to ensure synergy of regional efforts on monitoring of investment and results in water sector

The Session then formulated overall recommendations for submission to the Asia-Pacific Water Summit (APWS) for consideration, as follows.

- The various experiences in the region pointed to the urgent needs to improve effective policies for monitoring of investment and results, and to develop integrated mechanisms in most of the developing countries for better policy and decision making, as required by the prevailing rapid socio-economic growth in the region.
- The key KRA4 recommendations on follow-up action related to policies on Monitoring of Investment and Results were formulated as reported in the detailed Session Report for the following sectors: Water Supply and Sanitation; Agricultural Water Management; Ecosystem Management; Water-related Disasters Management; and also for Integrated Water Resources Management, for consideration of the APWS.

The Session agreed that monitoring of investment and results in water would include all aspects of water resources management. The Session noted the keen interest of countries in South Asia, East Asia and South-East Asia in creating a regional framework for regular consultation at the Ministerial level. In view of this fact, the initiative to expand the existing frameworks of subregional Ministerial Conference on Water toward the development of a Council of Ministers on Water Resources for Asia and the Pacific is therefore firmly founded in the ongoing regional efforts.

Messages

(a) The various experiences at the country and regional levels discussed at the 3rd Southeast Asia Water Forum and the Session pointed to the urgent needs to improve effective policies for monitoring of investment and results, and to develop integrated mechanisms in most of the developing countries for better policy and decision making, as required by the prevailing rapid socio-economic growth in the region. This fact is reflected in the gaps of effective linkage of monitoring data to socio-economic decision making; gaps in linking water resources management monitoring to decentralization; gaps in the legal framework for effective monitoring; and gaps in effective resources allocation. This fact also led to difficulties in embarking on major improvement in the software for water resources management.

(b) The water resources management issues and actions taken to deal with them are mainly at the local level. Past regional efforts usually faced with difficulties in bringing countries together, unless they aimed at the common goals such as the MDGs on water. Further regional efforts should be made to link findings of national efforts on monitoring of investment and results in water resources management to priority issues of the region, such as water supply and sanitation, water and climate change, water and food security and water and nature.

(c) Strengthening monitoring of investment and results in the region is expected to help APWF to sharp focus of regional cooperation and improve prioritization of joint activities.

Concrete Initiatives

Initiative Title

To develop programmes aiming at strengthening regional capacity in monitoring of investment and results for more effective cooperation through the Asia-Pacific Water Forum.

Specific Objectives

- To create knowledge hubs in water supply and sanitation system, ecosystem management, irrigation modernization and water-related disaster management to support countries in monitoring of investment and results.
- To convene a ministerial conference on monitoring investment and results with special attention to water supply and sanitation; water and climate change; water and food security; and water and nature.
- To support the existing frameworks of sub-regional Ministerial Conference on Water in monitoring of investment and results, starting with those related to MDGs in water, so as to expand these frameworks towards the development of a Council of Ministers on Water for Asia-Pacific.

Implementing Organizations

UNESCAP in its ongoing partnership with ADB and UNDP in cooperation with WHO and UNICEF is expected to expand the partnership to include other international organizations and UN bodies will be responsible for the coordination of efforts.

Active/Potential Partners at Sub-Regional Level/Country/Sub-Country

UNESCAP will focus its efforts within the framework of cooperation with APWF to support ASEAN countries in their commitment to accomplish MDGs in water by 2010. In addition, UNESCAP will also support other subregional initiatives related to monitoring of investment and results such as those of FAO-IWMI, IUCN-ARRN (Asian River Rehabilitation Network) and Typhoon Committee-ICHARM.

Implementing Schedule Before and After the 1st APWS

Prior to the 1st APWS, efforts of UNESCAP received support from many organizations such FAO, IWMI, WHO, UNICEF, the Typhoon Committee and the South-East Asian Water Partnership and the ASEAN Ministers on Water. Due to the fact that these efforts of UNESCAP and partners had not received strong support from many countries in the region at the 1st APWS, especially on the initiative to establish a Council of Ministers on Water for Asia-Pacific (COMWAP), it is now necessary to move slowly and carefully on this initiative, starting with subregional set up such as that for South-East Asia. It is now planned to support the South-East Asia Water Forum in its efforts to improve monitoring of investment and results at the fourth South-East Asian Water Forum to be held in the Philippines in 2009.

How to Assess the Performance/Progress

It is expected that regular reporting would be made as part of the monitoring of efforts on MDGs and also as part of the preparation for major meetings of APWF.

How will the Initiative be Funded and Sustained

As this initiative will build on existing initiatives and frameworks in the initial stages (to focus on MDGs on water), it is expected that no major financial resources would be required, except for activities aiming at capacity building in developing countries.

Regional Launch of the International Year of Sanitation (IYS) 2008

organized by Japan Bank for International Cooperation (JBIC), United Nations Economic and Social Commission for Asia and the Pacific (UNESCAP), United Nations Department of Economic and Social Affairs (UNDESA), United Nations Children's Fund (UNICEF), and World Health Organization (WHO) in cooperation with "Partner Group" United Nations Human Settlements Programme (UN-HABITAT), and Asian Development Bank (ADB)

Objective

- To enhance awareness of the public through the media of the decision of the United Nations General Assembly to declare 2008 as the International Year of Sanitation to mobilize commitment and support of all key stakeholders in its implementation;
- To mobilize commitment of countries in Asia-Pacific to work towards achieving the Millennium Development Goals in water and sanitation;
- To mobilize support from the private sector in the implementation of IYS-2008;
- To share experiences of the public sector, NGOs and key stakeholders in the implementation of MDGs on water and sanitation for replication.

Agenda

Welcome Statement by Mr. Koji Tanami, Governor, JBIC

Welcome Address by Mr. Masayoshi Namiki, Vice Minister of the Ministry of the Environment, Government of Japan

Inaugural address by His Royal Highness Prince Willem-Alexander of The Netherlands, Chairman of UNSGAB

Sanitation in the World and Asia-Pacific Region

CHAIR: Mr. Kazushi Hashimoto, Executive Director, JBIC

Report on the Current Status and Challenges on Sanitation in Asia-Pacific by Ms. Clarissa Brocklehurst, Coordinator of the UN-Water Task Force on Sanitation

Water Web Project on Google Map/Earth by H.E. Mr. Yoshiro Mori, President of Japan Water Forum, Former Prime Minister of Japan

Video film on sanitary situation in developing countries by the World Toilet Organization

Lessons Learned from Governments and the Public Sector

CHAIR: Ms. Clarissa Brocklehurst, Coordinator of the UN-Water Task Force on Sanitation

Reports from EASan and SACOSan by Dr. Twisuk Punpeng, Representative of East Asia Ministerial Conference on Sanitation and Hygiene 2007

Lessons learned from JBIC's experience in improving sanitation, Professor Tetsuya Kusuda

Experiences and struggles for expansion of sewerage systems in Bangkok, Mr. Chanchai Vitoonpanyaku, Deputy Director General, Bangkok Metropolitan Administration

Experiences of coping with sanitation problems in Guangxi, China, Mr. Lin Jiang, Secretary General of the Jiu San Society in Guangxi, China

Questions and Answers

Sanitation and the Private Sector

CHAIR: Mr. Andre Dzikus on behalf of Mrs. Anna Tibaijuka, Executive Director, UN-Habitat

Keynote Speech on "Financing Sanitation – Looking for New Approaches" by Ms. Barbara E. Evans, EASAN Conference Director and International Consultant on Water and Sanitation

Panel Discussion on "Sanitation as Investment opportunities: Perspectives of the Private Sector"

PANELISTS:

- Ms. Barbara E. Evans, EASAN Conference Director and International Consultant on Water and Sanitation
- Ms. Clarissa Brocklehurst, Coordinator of the UN-Water Task Force on Sanitation
- Dr. K. E. Seetharam, ADB
- Mr. Gerard Payen, Aquafed

- Dr. Twisuk Punpeng, Thailand
- Mr. Jack Sim, WTO

Sanitation and the Civil Societies

CHAIR: Dr. Rivka Kfir, Chairperson of the Stream of Knowledge

Scaling Up the Solution Together: Strategies, Leverage, Synergies: Plan, achievements and strategies of the World Toilet Organization, Jack Sim. Founder World Toilet Organization

Panel Discussion on "Partnership on Sustainable Sanitation"

PANELISTS:

- Mr. Jack Sim, WTO
- Ms. Lee Kyung-Hee, WTA
- Mr. Henry Northover, Water Aid
- Ms. Rosario Villaluna, Stream of Knowledge
- Mr. Quamrul Islam Siddique, President, Bangladesh Water Partnership

Concluding Session

MODERATOR: Dr. Ken Chen, WHO Representative to the Pacific

Concluding Statement by Mr. Rae Kwon Chung, Director, Environment and Sustainable Development Division, UNESCAP

Summary of Discussions

The Session benefited from the participation of H.E. Mr. Masayoshi Namiki, Vice Minister of the Environment of Japan, Mr. Koji Tanami, the Governor of JBIC, and particularly H.R.H. Crown Prince of Orange who is currently the Chair of the United Nations Secretary-General's Advisory Board of Water and Sanitation and H.E. Former Prime Minister Yoshiro Mori, President of the Asia-Pacific Water Forum in the opening segment of the Session. Their participation has accorded top-level attention of the public and media as well as participants of the APWS. The launching by H.E. President Yoshiro Mori of the Water Web Project on Google Map/Earth was recognized as an important milestone for not only regional but also global efforts in carrying out activities of IYS.

After the opening segment, the Session devoted nearly three hours to discuss four key groups of issues and experiences on the challenges in Asia-Pacific in sanitation, experiences in the public sector on sanitation, and roles of the private sector and civil societies in the promotion of sanitation.

The Session also referred to the earlier efforts and recommendations of the Ministerial Conference on Sanitation of South Asian Countries held in 2006 and the back-to-back East Asian Ministerial Conference on Sanitation and Hygiene held in Beppu on 30 November-1 December 2007 so as to ensure best outcomes for the regional efforts in the implementation of the International Year of Sanitation 2008.

The Regional Launching pointed out the fact that if the MDGs are to be achieved, 2.4 *billion* people need to gain access to sustainable sanitation services, or 96 million a year between 1990 and 2015. Countries will need to see that enough wisdom, will, resources and creativity are invested to make such a huge change possible. It also pointed out the need for Governments and individuals to understand the nature, magnitude and urgency of the sanitation crisis. It also pointed out that poor sanitation not only pushes already disadvantaged sections of society into deeper illness, poverty and indignity, which is also a key delimiting factor to sustained economic growth.

The Session on the Regional Launching provided opportunities to exchange experiences for more effective contribution of the public and private sectors and civil societies as well as partnerships towards achievements of MDGs in water and sanitation, for which many participants from various countries in the region as well as international organizations contributed their experiences and good practices. These experiences also called for more opportunities for the promotion of investments from all key stakeholders to promote a rapid up-scaling of progress in access to sanitation. This confirms the need to call Governments in the region to also launch the International Year of Sanitation in their respective countries.

The Session reiterated the following two most important recommendations, which built on the findings and recommendations of the Declarations of SACOSAN and EASAN, and submitted these for the Summit for consideration:

1. The Asia-Pacific Water Forum recognize SACOSAN 2006 and EASAN 2007 and their respective follow-up as an integral part of the APWF process; and
2. Create a regional platform for cooperation in sanitation and hygiene, to include a regular Ministerial Conference on Sanitation and Hygiene at two-yearly intervals.

Messages

- Asia-Pacific Water Summit commits to achieve MDGs in sanitation

- Sanitation is an integral part of water resources management and its development must be mainstreamed into the integrated water resources management process
- Sanitation as an economic investment
- Strengthening mechanism on monitoring of investment and results in sanitation is an essential part to support the Asia-Pacific Water Summit to achieve MDGs in sanitation

Concrete Initiatives

Initiative Title

(a) Launching of the IYS 2008 in Asia & the Pacific
(b) Launching of the "Water Web Project on Google map & earth"

Specific Objectives

- To mobilize commitment of Governments and financing institutions to ensure that sanitation is included in the national development agenda, starting with the adoption of 2008 as an International Year of Sanitation
- To devise measures to promote action-oriented programmes as follow-up of subregional conferences on sanitation, including SACOSAN and EASAN
- To strengthen the current mechanism on monitoring of sanitation targets in the region through enhancing policy making to monitoring results
- To enhance participation of the private sector and civil societies in sanitation development

Implementing Organizations

(a) UNESCAP, WHO, UNICEF and WSP of the World Bank in cooperation with ADB, the World Toilet Organization and World Toilet Association
(b) Japan Water Forum

a. *Active/Potential Partners at Sub-Regional Level/Country/Sub-Country*

Network of WHO, UNICEF and WSP of the World Bank and country focal points on water supply and sanitation

b. *Implementing Schedule Before and After the 1st APWS*

(a) WHO, UNICEF and WSP of the World Bank organized EASAN on 30 Nov-1 Dec 2007; WHO, UNICEF, WSP of the World Bank, UNESCAP and partners to commemorate the World Water Day with focus on IYS; to assist countries in holding the 3rd SACOSAN in 2008 and the 2nd EASAN in 2009; to mobilize participation of developing countries in launching IYS-2008 in the national level; and to organize regional meeting on policy changes related to IYS and progress in MDGs on sanitation in 2008

(b) JWF launched the Water Web Project on Google Map/Earth at APWS

c. *How to Assess the Performance/Progress*

- Number of countries adopt IYS-2008
- Number of projects/actions implemented by countries at the subregional level: SACOSAN, EASAN

d. *How will the Initiative be Funded and Sustained*

- Most of related activities are expected to be funded as part of ongoing activities
- Additional resources are required to hold the third SACOSAN, the second EASAN and the regional meeting on policy changes related to IYS and progress in MDGs on sanitation in 2008.

The Small Islands' Dialogue on Water and Climate

organized by Pacific Islands Applied Geoscience Commission (SOPAC) and International Centre for Water Hazard and Risk Management (ICHARM)

Objective

Call attention to the vulnerability and particular needs of small island countries to effectively manage their water resources under a highly variable and changing climate.

Agenda

Opening Remarks by the Moderator, Ms. Cristelle Pratt, Director, SOPAC

Video on "Islands and Climate Change" introduced by H.E. Mr. Anote Tong, President Kiribati

Roundtable Panel Discussion

MODERATOR: Ms. Cristelle Pratt, Director, SOPAC

PANELISTS:

- H.E. Mr. Anote Tong, President of the Republic of Kiribati
- Hon. Mr. Tuisugaletaua Sofara Aveau, Minister for Works, Samoa
- Dr. Chen Ken, Head South Pacific Regional Office, WHO
- Mr. Latu Kupa, Executive Director, Pacific Water Association, PWA
- Dr. Kuniyoshi Takeuchi, Director, ICHARM
- Mr. Salvano Briceno, Director, ISDR

Leaders' Commitments

Wrap-up by the Moderator

Summary of Discussions

The organisers of the 1st Asia Pacific Water Summit were thanked for recognising the need for a small islands dialogue in respect of water and climate and including a special session focusing on Small Islands at the Summit.

The session sought to provide guidance and commitment to supporting Small Island Developing States of Asia and the Pacific Region to actively and effectively pursue continued implementation of water and sanitation initiatives through strengthened and innovative partnerships at local, national, regional and international levels.

The large participation by Pacific Heads of State at this Summit was a testament of their strong political commitment to meeting future water challenges and their efforts to cope with an increasingly variable climate, and adapt to the future effects of global climate change.

There was strong reinforcement that there are clear and inextricable links between water and climate, ocean and climate and the critical relationship between the interface of island and ocean interactions.

There was also recognition that there is a real need to remain mindful of the cross cutting nature of climate and the need to look at various other sectors such as health and education within the space, when tackling the development challenges related to water and adaptation to climate change, variability and extreme weather events.

It was contended that without political commitment and a strategic policy framework within which to operate the challenge to provide safe drinking water and adequate sanitation is that much more difficult and therefore the main challenge is to get the political commitment and then to maintain the political profile and commitment to ensure that momentum is not lost.

Excellent examples were shared of various national initiatives that are being implemented through a range of proactive measures and options and it was evident that from these there is and has been some very good progress and increased momentum in implementation over the past few years; such as Phases I and II of the Kiribati Adaptation Project and Samoa's National Adaptation Plan for Water.

However, it was also widely conceded and accepted that one-size-"does-not"-fit all and that various options and arrangements need to be examined for different situations.

The Session was able to provide the space to achieve the objectives of the APWF to provide countries and organisations in the Asia Pacific region with a common platform and voice to articulate the region's strategies and promote its achievements in solving water problems and to showcase leadership in decision-making, excellence in practice and innovation and results that have had a substantial impact at the grassroots level.

Attention was raised to the opportunity that presents itself at this moment, to mainstream Climate Adaptation, Disaster Risk Reduction and Water Safety Planning into Integrated Water Resources Management. Combined with adequate priority given to water and sanitation in National development plans and strategies, these measures will be the best approaches to achieve the MDG targets of halving the proportion of people without access to safe drinking water and basic sanitation by 2015 and be prepared for the future for our peoples to live free and worthwhile lives.

Messages

Integrated approaches are needed for water resources management and in fact that similar approaches are required if we are to effectively tackle disaster risk reduction across the various other development sectors; and in this respect suggested that there has and must be a paradigm shift from disaster response to water-related disasters to disaster risk reduction and disaster management.

In order to continue to make the positive advances that are being made within the small island States of the Asia and Pacific Region it was agreed that there is a need for continued cooperation and collaboration at all levels and through various partnership modalities. The Session sighted the GEF-Pacific Alliance for Sustainability, the establishment of robust Early Warning Systems and the upcoming Climate negotiations in Bali as immediate opportunities to link to and explore in respect of addressing the challenges related to water and adaptation to climate change, variability and extreme weather events.

Concrete Initiatives

Initiative Title: Pacific Partnership Initiative on Sustainable Water Management. http://www.sopac.org/tiki/tiki-index.php?page=CLP+Pacific+Partnership

Specific Objectives

Mainstream Climate Adaptation, Disaster Risk Reduction and Water Safety Planning into Integrated Water Resources Management linked to National development plans and strategies as best approach to manage the impacts of climate change relevant to water and sanitation.

Implementing Organizations and Active/Potential Partners

USP, SOPAC, SPC, SPREP, Pacific Islands Forum Secretariat, Fiji School of Medicine, AusAID, ADB, EU, GEF, NZAID, UNDP, UNEP, UNESCO, WHO, WMO, UNICEF, OXFAM, WERI, WSSCC, UNU, ANU, NIWA, BOM, IWC, IFRC, CSIRO, FSPI, LLEE, WWF, GWA, GPA, GWP, PWA (and others)

Implementing Schedule Before and After the 1st APWS

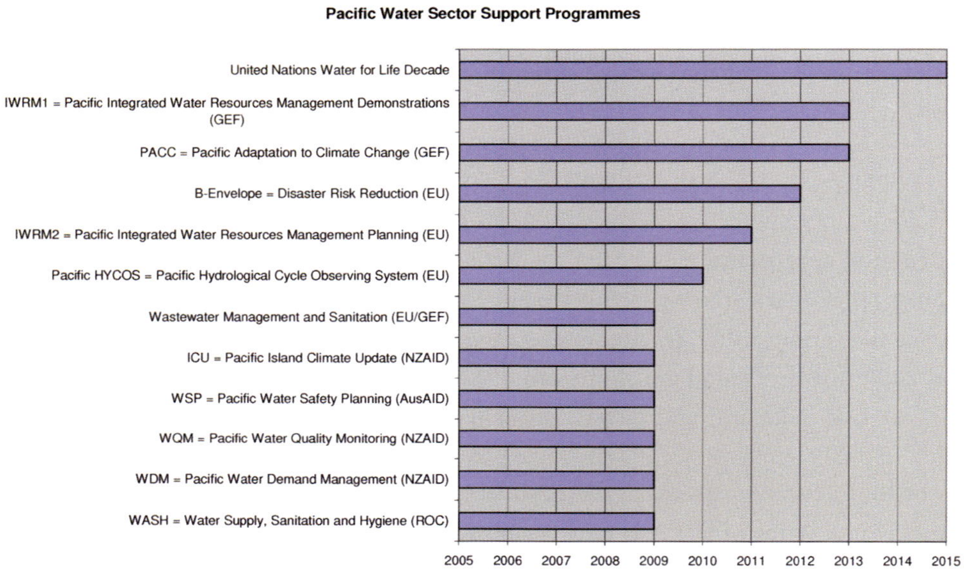

How to Assess the Performance/Progress

Pacific Water Action Matrix

The Pacific Partnership Initiative on Sustainable Water Management is coordinated and facilitated by the intergovernmental organisation SOPAC (Pacific Islands Applied Geoscience Commission), which has the regional mandate to support the Pacific Island Countries in water & sanitation. SOPAC is assisted in this task by a deputy facilitator, the University of the South Pacific (USP).

The Partnership has a Facilitator (based at SOPAC) who is responsible for implementing the core functions of the partnership: liaising between the regional stakeholder groups and their sub-networks; researching and receiving stakeholder information on on-going and planned water activities; tracking donor and development agency programmes; identifying areas requiring implementation; and coordinating proposal submissions and project implementation. The Facilitator is also responsible for high-level advocacy of the strategic approach.

Monitoring and evaluation are carried out using a matrix inventory of previous, existing, planned and proposed activities, including details of the stakeholders involved, the intervention objectives, implementation duration and status, and anticipated impact.

http://www.sopac.org/tiki/tiki-sopac_actionmatrix.php?

A Monitoring and Evaluation Framework with IWRM, Water Use Efficiency (WUE) and Water Safety Planning (WSP) indicators will be established under the GEF and EU Water Facility funded Pacific IWRM programme. The framework will include process indicators, environmental and socio-economic stress reduction indicators with linkages to the monitoring of the MDG targets on water and sanitation.

The Facilitator enables countries and development agencies to: identify successful previous activities and therefore improve the sustainability of subsequent interventions; reduce and prevent duplication of activities; link country requirements to development programmes (and vice versa); and augment existing and proposed activities nationally and regionally.

The various interventions in the water and sanitation sector are or will be imbedded in National sustainable development plans, water sector strategies, policies and action plans and will be incorporated specifically in National IWRM plans which are to be developed over the period 2008–2009.

How will the Initiative be Funded and Sustained

Individual Pacific Island Countries' government contributions to national action implementation will contribute to achieving the objectives of the initiative with support provided by various donor organizations and international agencies: ADB; AusAID; NZAID; NZODA; DFID; JICA; GEF; USAID; EU; Taiwan-ROC.

Commitment for Ensuring Water Security in the Aral Sea Basin – Cooperation versus Competition

organized by International Fund for saving the Aral Sea

Objective

- Discussion of the existing problems of the Aral Sea Basin in the context of APWF Priority Themes and proposal of new initiatives to achieve sustainable water management in Central Asia.
- Attraction of attention of the world community including international organizations and donor agencies to the Aral Sea basin problems as well as their solution.
- Mobilization of efforts of the Central Asian countries, UN institutions, international organizations, donor agencies, NGOs etc. for achievement of MDGs in Central Asia.

Agenda

Opening Remarks by the Moderator, Mr. Loic Fauchon, President of WWC

Keynote Speech by H.E. Mr. Emomali Rahmon, President of the International Fund for the Aral Saving, President of the Republic of Tajikistan

Speeches by Head of Official Delegations

- Mr. Anatoliy Ryabtsev, Chair of the Water Resources Committee, Ministry of Agriculture of the Republic of Kazakhstan
- Mr. Baratali Koshmatov, Director General of the Water Economy Department of the Ministry of Agriculture, Water Economy and Processing Industry of the Kyrgyz Republic
- H.E. Mr. Said Yoqubzod, Minister of Melioration and Water Resources of the Republic of Tajikistan
- Mr. Shavkat Khamraev, Deputy Minister of Agriculture and Water Economy of the Republic of Uzbekistan
- H.E. Mr Seyid Baghirov, Minister of Ecology and Natural Resources of the Republic of Azerbaijan
- Mr. Sulton Rahimov, Chairman of the Executive Committee of the International Fund for saving the Aral Sea

Questions & Answers

Remarks by Donors and International Organizations' Representatives (ADB, UNESCO, INISDR, WMO, UNESCAP, JICA, GWP, JWF, ICID)

Conclusion by Session Reporter, Mr. Andras Szollosi-Nagy, Assistant Director General, UNESCO

Summary of Discussions

The key word of the session was COOPERATION. "We need, we must, we are ready to cooperate!" - this is a wonderful message that came out. But all participants express an opinion that without support of the world community it will be very difficult to solve the Aral Sea Basin problems. That's why the President of IFAS, President of the Republic of Tajikistan H.E. Mr. Emomali Rahmon giving his view on this issue once again calls international organizations and donor agencies to assist Central Asian countries on solution of these problems and achievement of MDGs. H.E. Mr. Emomali Rahmon has also expressed an anxiety on worsening of the problems by global climate change, population growth and rapid economic development and proposed some new initiatives to contribute to improvement of water resources management in the Aral Sea Basin countries. In particular H.E. has invited interested parties to contribute on creation of the International Consortium on using of the fresh water resources of the Sarez Lake in Tajikistan for drinking water supply of millions people in Central Asia thereby reducing its break risk as well as to gather at the end of June 2008 in Dushanbe to discuss water related disaster reduction within the framework of the International Conference to be organized by the Government of Tajikistan.

High-ranking officials from Central Asia countries and representatives of the international organizations have also noted the existing problems in the region and pinned their hopes that the 1st Asia-Pacific Water Summit will become a new reference point on mobilization of efforts of the Central-Asia countries, the United Nations institutions, international organizations, donor-countries, NGOs, private sector and civil society in solution of the Aral Sea Basin problems as well as stirring up of the wide international support to the Central Asian countries.

All session participants have expressed hopes that the Summit will promote development of initiatives of the Heads of the Central Asian states on giving to the International Fund for saving the Aral Sea the status of the United Nations institution for coordination of activity of the international organizations and donors with announcement of the Aral Sea Basin as the priority pilot region for achievement of the Millennium Development Goals as well as new initiatives made by the President of IFAS, President of the Republic of Tajikistan H.E. Mr. Emomali Rahmon during the session.

Messages

Water is common for humanity, which have fundamental importance for life conservancy on the Earth and sustainable development of the present and future generations and therefore it should be recognized as most valuable resource. Everybody should understand this point and make concrete input on implementation of tasks of the International Decade for Action «Water for life», 2005–2015, announced to make better water situation in the world.

Taking into account worsening situation with drying of the Aral Sea, which is one of the biggest disasters in the Asia-Pacific region, and increasing problem of water resources use in its basin, caused by climate change and population growth to call the world community to assist Central Asian countries to solve the Aral Sea Basin problems for population well-fare and achievement of MDGs as well as to render an assistance to the International Fund for saving the Aral Sea to implement the Program of concrete actions on improvement of ecological and socio-economical situations in the Aral Sea Basin for the period of 2003–2010 (ASBP-2).

To welcome and support a proposal of Tajikistan on organization of the International Conference on Water Related Disaster Reduction at the end of June 2008 in Dushanbe. To invite APWF to announce the outcomes of the 1st Asia Pacific Water Summit in this important water event as well as APWF members and other stakeholders to attend it.

The Governments of the Central Asian countries will undertake commitments to give priority importance to water issues on preparation of the national and regional development plans and strategies, to strengthen cooperation to ensure water security in the Aral-Sea Basin countries and to increase contribution of all stakeholders to integrated water resources management in Central Asia sub-region.

Concrete Initiatives

Initiative Title

- to give the status of the United Nations institution to the International Fund for saving the Aral Sea for coordination of activity of the international organizations and donors with announcement of the Aral Sea Basin as priority pilot region for achievement of the Millennium Development Goals;

- to set up an International Consortium on using of the fresh water resources of the Sarez Lake for drinking water supply in Central Asia thereby reducing its break risk.
- to create APWF centers in all five sub-regions for strengthening of Forum's role and improvement of water resources management as well as creation of APWF Central Asian sub-regional center in Dushanbe;
- to develop Water Strategy of the Central Asia sub-region in context of the APWF Priority Themes with concrete commitments on water financing, disaster management in interrelation with ecosystem approach in development in collaboration with APWF.
- to create the Aral Sea Basin Public Water Council.

Specific Objectives

- To increase public awareness about the Aral Sea crisis and existing problems in its basin;
- To mobilize efforts of the Central-Asia countries, the United Nations institutions, international organizations, donor-countries, NGOs, private sector and civil society to solve the Aral Sea Basin problems as well as stir up of the wide international support to sub-region countries, impacted by this crisis;
- To enhance transboundary collaboration through more open and informed dialogue;
- To contribute to sustainable water management in order to achieve the targets of MDGs in all sub-regional countries by 2015;
- To enhance coordination at the sub-regional and national levels by increasing stakeholders representation in the Aral Sea Basin Public Water Council

Implementing Organizations

- Executive Committee of the International Fund for the Aral Sea Saving (EC IFAS)
- Interstate Commission for Water Coordination in Central Asia (ICWC)
- Interstate Commission for Sustainable Development (ICSD)
- GWP CACENA

Active/Potential Partners at Sub-Regional Level/Country/Sub-Country

National Water Authorities (Ministries, Departments, Agencies), Basin Water Organizations (transboundary), APWF, Regional offices of international organizations and donor agencies (ADB, WB, SDC, UNDP, UNESCO, UNISDR, etc.)

Implementing Schedule Before and After the 1st APWS

- April 2007 – the 1st Sub-regional workshop towards APWS (Almaty, Kazakhstan)
- May-September 2007 – sub-regional consultations and drafting of the "Provisions for Sub-Regional Conception for Central Asia on Priority Themes of APWF"
- September 2007 – the 2d Sub-regional workshop towards APWS (Turkmenbashi, Turkmenistan)
- 5–7 October 2007 – Meeting of the Head of States of Central Asia in Dushanbe, Tajikistan
- November 2007 – Approval of "Prerequisites for Conception of Central Asia sub-region in the context of APWF Priority Themes" by IFAS Board
- December 2007 – Panel session "Commitment for Ensuring Water Security in the Aral Sea Basin – Cooperation versus Competition" during 1st APWS in Beppu, Japan
- June, 2008 – International Conference on water-related disaster reduction (Dushanbe, Tajikistan) – presentation of the outcomes of the 1st APWS
- 2008 – Follow up activities of the Aral Sea Basin Public Water Council
- 2008 – Preparation process for the 5th World Water Forum, Istanbul - March 2009.

How to Assess the Performance/Progress

There is expected to make a joint message to world community to assist Central Asian countries to solve the Aral Sea Basin problems as well as to obtain commitment from all sub-regional governments to support agreed goals and tasks of the new "Strategy" for the Aral Sea Basin.

Achieved understanding of the importance of the co-ordination at all levels of water management hierarchy, and of the input of each participant into integrating water resources management.

How will the Initiative be Funded and Sustained

There should be agreed procedures and methods for stabilizing water supply, providing equitable water distribution, and establishing the proper public control by water users themselves. At the same time, the technical and financial assistance of the countries and local governments is necessary. The proper contribution from international donor agencies is expected.

Water-related Disaster Management: Priority Theme B

organized by International Centre for Water Hazard and Risk Management (ICHARM)

Objective

The Asia-Pacific is the most vulnerable region to water-related disasters in the world that hinder sustainable development and poverty reduction. Between 1960 and 2006, over 600 thousand casualties were recorded that account for over 80 percent of casualties worldwide, in addition to US$ 8 billion worth of economic damage during the same period.

Under this situation, the objective of the session on "Water-related Disaster Management" aims to create a new momentum for commitment and leadership to strengthen and ensure the effectiveness of our efforts to achieve a positive change in the way water-related disasters is managed in the Asia-Pacific region.

Agenda

Opening Remarks, Introduction of the Draft Recommendation by Dr. Kuniyoshi TAKEUCHI, Director, ICHARM

Speech by Mr. Junzo YAMAMOTO, Parliamentary Secretary, Ministry of Land, Infrastructure and Transport, Japan

Speech by H.E. Mr. Saif-ud-din Soz, Minister of Water Resources, India

Speech by H.E. Mr. HU Siyi, Vice Minister of Water Resources, People's Republic of China

Presentation by Dr. Andras Szhollosi-NAGY, Assistant Deputy Director General, UNESCO

Presentation by Mr. Avinash TYAGI, Director, Hydrology and Water Resources Department, WMO

Presentation by Mr. Salvano BRICENO, Director, UN/International Strategy for Disaster Reduction (ISDR)

Presentation by Mr. Wouter Lincklaen ARRIENS, Lead Water Resources Specialist, ADB

Presentation by Dr. Katumi MUSIAKE, Professor, Fukushima University

Discussion

Adoption of Messages

Closure

Summary of Discussions

High-level policymakers, guests from international organizations, academic expert and government officers discussed widely on the water-related disaster management. Speeches/presentations were made to cover various aspects of water-related disaster management. The common issues included was the anticipated Climate Change and how to take flexible adaptive measures to realize the integrated approach towards water-related disaster management towards better water-related disaster management.

Through the discussion, the following views were shared

In the Asia-Pacific region, even under current geographical, meteorological and socio-economic conditions, the level of Water-related Disaster Risk Reduction (DRR) Management is far from sufficient and many vulnerability factors are ever-increasing. As reported by the IPCC, it is now evident that the risks resulting from Climate Change may cause extra burdens such as extreme precipitation and tropical cyclones, more frequent droughts, enlarged tsunamis/storm surges. Countries need to share knowledge and resources with the people with adequate information and raise awareness to cope with threats from Climate Change.

Water-related DRR is the main means of adaptation to Climate Change, which is urgent and necessary. Each country should effectively combine both structural and non-structural measures, depending on its situations, to reduce the risks from water-related disasters, thus contributing to the promotion of water-related DRR to protect people, properties and well-being. While non-structural measures should be more effectively adopted, structural measures will continue to play an important role to improve the protection level from expected risks, for which resources need to be ensured for steady implementation of necessary measures.

From the viewpoint of protecting human lives, it is necessary to circumvent risks by ensuring timely and smooth evacuation actions at the time of disasters, which will be provided by measures such as hazard maps and early warning systems. Furthermore, adaptive measures should be taken in socially acceptable manner, which have sustainable development perspectives depending on local conditions. For this purpose, governments should continue to bear the overall responsibilities on nation-wide policy planning, mobilizing resources and implementing measures in cooperation with local governments and relevant agencies.

As agreed by the Ministers during "Ministers' Forum on Asia-Pacific Infrastructure Development" on 30 August, 2007 in Beijing, water-related DRR needs to be realized within the context of Integrated Water Resources Management (IWRM) which includes proper management of excessive as well as scarce water. Integration needs to reflect various social, economic and cultural situations in the region. Water-related DRR in the IWRM perspective needs to be incorporated, with the highest priority, into national development plans with clear goals/targets. Efforts need to be made to establish a standardized procedure for disaster preparedness to monitor and better manage water-related disasters.

Quantitative effects of Climate Change to hydro-meteorological conditions will be gradually and carefully gathered to formulate adaptive strategies thus enabling actions be taken efficiently on time. The Asia-Pacific region should share experiences, methodologies, technologies and resources and help each other to comprehensively and properly promote water-related disaster management where willingness for contribution was expressed by Japan. It is also important to create a knowledge sharing platform (knowledge hub) and to endeavor to foster experts and enhance institutional capacities based on local needs. It should also be stressed that women should play active roles in the course of water-related disaster management.

In closing the session participants adopted following key messages towards better management of water-related disasters, that were shown in the Policy Brief paper at the beginning of the Summit.

Messages

- Integrate water-related Disaster Risk Reduction (DRR) into national development plans, recognizing adaptation to increasing risks from climate change as a "highest" priority issue
- Recognize the importance of IWRM for water-related DRR and the need to strengthen comprehensive structural and non-structural measures

- Establish national and local goals/targets for water-related DRR, taking the impacts of climate change into consideration
- Develop preparedness indices for water-related DRR for the Asia-Pacific region
- Develop water-related disaster warning systems and human capacities

Concrete Initiatives

Initiative Title

Establishment of regional Knowledge Hub for Water-related Disaster Risk Reduction

Specific Objectives

Establishment of a Center of Excellence to comprehensively provide following services for water-related disaster risk reduction in Asia-Pacific region

- Capacity building (Implementation of master-course level academic education for disaster management specialists)
- Flood warning/alert systems (Development of technological basis that enables provision of flood forecasting for specific river basins utilizing satellite-based data)
- Information sharing and dissemination (issuance of periodical newsletters, large flood reports, etc)
- Development of preparedness indices that enables to monitor progress of water-related DRR in each country/local governments

Implementing Organizations

The International Centre for Water Hazard and Risk Management (ICHARM), Public Works Research Institute (PWRI), Japan

Active/Potential Partners at Sub-Regional Level/Country/Sub-Country

ADB, UNESCO Category II Centers, etc.

Implementing Schedule Before and After the 1st APWS

- The Master-level academic training course (1-year) started from Oct. 2007. The course will be continuously organized
- Periodical newsletters have been issued and large flood reports will be soon be published on periodical manner
- Indices to monitor preparedness for water-related disasters will be developed and proposed (possibly at 5th WWF in March 2009)
- Technological base for satellite-based flood warning/alert system will be developed and applied to specific basins in the AP region

How to Assess the Performance/Progress

- The number of trainees who completed training courses
- The number of publications/newsletters issued
- The number of countries who applied proposed water-related disaster preparedness indices & monitored their progress in water-related DRR
- The number of river basins and/or flood management organizations that are provided with satellite-based flood warning systems

How will the Initiative be Funded and Sustained

ICHARM funds will be provided to fulfill basic requirements for listed activities. Financial support by JICA has been provided for implementing master-level water-related disaster management training. The collaboration with other agencies such as ADB, is expected to significantly enrich and accelerate the process.

Water for Development and Ecosystems: Priority Theme C

organized by The World Conservation Union (IUCN) and Food and Agriculture Organization of the United Nations (FAO)

Objective

The objective of this theme was to secure water for long-term, multi-sectoral needs, generating both development and environmental dividends.

The goal was to present and confirm support for the policy recommendations and to launch Asia Pacific follow-up initiatives.

Agenda

Welcome Remarks by the Session Chair, Dr. Changchui He, FAO

Welcome Remarks by the Session Chair, Ms. Julia Marton Lefevre, IUCN

Keynote Address by Honorable Mr. Gyanendra Bahadur Karki, Minister of State for Water Resources, Nepal

Presentation of the Recommendations for Theme C from the Asia-Pacific Water Forum Policy Brief 2007 by Dr. Ger Bergkamp, Head, Global Water Programme IUCN

Panel Discussion

MODERATOR: Dr. Donald Blackmore, Chairman, eWater Cooperative Research Centre

PANELISTS:

- Honorable Mr. Gyanendra Bahadur Karki, Minister of State for Water Resources, Nepal
- Hon. Mahendra Pal Chaudhry, Minister of Finance, National Planning and Public Enterprises, Fiji
- Mr. Wu Shiqiang, Director, Nanjiang Hydrualic Research Institute, China
- Mr. Mukuteswara Gopalakrishnan, Secretary General, International Commission on Irrigation and Drainage

- Dr. Changchui He, FAO
- Ms. Julia Marton-Lefevre, IUCN
- Professor Tsugihiro Watanabe, National Institute for the Humanities, Research Institute for Humanity and Nature

Closing Remarks by Ms. Margaret Cately-Carlson, Chair, Global Water Partnership

Summary of Discussions

Mr. Changchui He (FAO) opened the session with a reminder of the MDG and World Food Summit targets of halving the number of undernourished in the region by 2015. He stressed that irrigated agriculture is the major water consumer and backbone of food security in the region, thus development and environmental targets will not be met if this sector does not adopt forward-looking strategies.

In her welcome comments, Ms. Julia Marton-Lefevre (IUCN) stressed that in order to create sustainable systems for development; water must be saved for the environment. She underlined the need for more integrated thinking to recognise the environment as a stakeholder in water-related decisions.

In his key note address Hon. Gyanendra Bahadur Karki highlighted the applicability of the session's recommendations to the situation in Nepal. He stressed that Nepal wanted to adopt the practice of Integrated Water Resources Management but that there was a lack of experience and knowledge of this approach in Nepal. In his remarks he also urged for a more meaningful dialogue between riparian countries, stressing the prevalence of transboundary river basins in Nepal.

Dr. Ger Bergkamp presented the Theme C policy recommendations (see policy document for specific recommendations) and there was general agreement of them among the keynote presenter and session panellists.

During the panel, a number of Ministers and experts presented the following key points:

Hon. Mahendra Pal Chaudhry underlined the importance of investing in ecosystems for development. He indicated his support for the FAO and IUCN initiatives launched at the session, but stressed that further advice was needed on how to tailor these to small island nations. He also reminded the session of the economic and natural vulnerability of small island nations such as Fiji, indicating that they are very much dependent on financial help and capacity building from others.

Mr. Wu Shiqiang commented on some of the problems associated with the use of water for agriculture in China, indicated fundamental principles of national policy: harmony between man and nature and development of a water-saving society. He underlined efforts to restore key ecosystems in the Karim and Balck Rivers, indicated that significant water-savings in irrigation had allowed a substantial reduction in relative water allocation to agriculture while continuing its development, and that China was planning to maintain national food security in the future without increasing water allocation to agriculture.

Mr. Mukuteswara Gopalakrishnan reaffirmed ICID's role in managing water for sustainable agriculture, highlighting the huge challenge faced in doubling the food production for the future needs of 8 million people. Mr. Gopalakrishnan underlined ICID's support for Theme C and the FAO-IWMI follow up initiative and it's commitment to work with others on the outcomes of the Session. He outlined particular areas were ICID could lend support as: modernizing irrigation system management, capacity building, benchmarking, assessing multiples roles of agriculture and climate change.

Dr. Changchui He (FAO) unveiled a joint initiative with the International Water Management Institute (IWMI) to re-invent irrigation and agricultural water governance in the Asia Pacific to meet the Millennium Development Goals.

Ms. Julia Marton-Lefevre stressed IUCN's commitment to establishing a community of practice of diverse stakeholders to implement environmental flows tools and applications in the Asia Pacific region as a follow up initiative to the Summit.

Professor Tsugihiro Watanabe, on behalf of the International Network for Water and Ecosystems in Paddy Fields (INWEPF), delivered a message to the Summit highlighting the extensive services related to society, culture and ecosystems provided by rice paddy systems and identifying three key areas for action: 1) Improvement of irrigation facilities and water management for food security and poverty alleviation; 2) Recognition and evaluation of multiple functions of paddy farming systems and their incorporation into water resources management; 3) Establishment of new on demand driven and service-orientated participatory management strategies.

A representative from the audience stressed that in order to mitigate and address adaptation to climate change issues, water-bourn transport can play a role as it is energy efficient and has low emission of carbon dioxides.

In her closing remarks, Ms. Margaret Catley-Carlson observed that the challenges to implementing recommendations remain formidable. She stressed that complex political work is needed to address the major water issues facing Asia-Pacific today. This is not simply a technical job. The full range of difficult governance issues need to be addressed if further progress is to be made. She also stressed that complex science and knowledge have a real role to play in addressing some of the challenges.

Messages

The session reaffirmed rigorous actions for securing the water needs of both development and the environment in the region, highlighting the need to invest in ecosystems as development infrastructure which must be maintained, restored, monitored and managed. The session also reiterated the region's dependence on water in order to achieve future food security and underlined the threat of climate change to this dependency. Irrigation and agriculture were therefore highlighted as key areas of reform, not only to increase food security, but also to alleviate poverty and improve the health of the environment. The session also stressed the importance of governance, in the form of transparent, inclusive and accountable decision-making to ensure that benefits of water management are shared more equitably across society.

The various speeches made it clear that much progress has been made in this sector recently and that there is now a common understanding concerning the way forward. However, participants stressed a sense of urgency in taking concrete action towards integrated thinking in which development enhances sustainability instead of threatening it. Panel speakers asserted that a mix of technical innovations and institutional reforms must form the basis for new strategies to manage agricultural water and achieve sustainable ecosystems and poverty alleviation. While the actual implementation of the Theme C policy recommendations presents a huge challenge, the initiatives launched by FAO and IUCN at the session form a robust framework for addressing these recommendations.

Concrete Initiative 1 - launched by FAO and IWMI

Initiative Title

Re-inventing irrigation and agricultural water governance in the Asia Pacific to meet the Millennium Development Goals: A regional programme to support action at all levels through knowledge sharing, capacity building and implementation.

Specific Objectives

The objective of the initiative is to support feasible pathways to re-invent irrigation and re-think agricultural water governance to ensure the social, economic and environmental outcomes that will ensure sustained food security and poverty alleviation in the river basins of the region. This will be achieved through the adoption of effective policies and strategies; through investment and modernization of all types of irrigation systems; and through improved agricultural water governance.

Implementing Organizations

FAO, IWMI

Active/Potential Partners at Sub-Regional Level/Country/Sub-Country

UNESCO-IHE; PUB, Singapore; Global Water Partnership; IPTRID; ADB; AIT; ICID; WBI; INWEPF; ESCAP; IUCN; Krishna CP, KNNL and MWR, Karnataka State, India; MWR, NCIDD, Shanxi DWR and JMK project, China; DID, MADA, Malaysia; MRCS and national MRC Committees; national ICID committees; CSIRO; NARBO; University of Melbourne; Vietnam: Cau RBO, MARD, MONRE, VIWRR; Indonesia: Citarum basin, BAPPENAS, etc. Pakistan: ARC, etc.

Implementing Schedule Before and After the 1st APWS

- Firming up of concept note and partnerships, August–November 2007
- Launching of the Initiative: APWS 3–4 December 2007
- Development, formulation of the initiative's management setup, confirmation of partnerships, detailed formulation of regional and river basin action components and resource mobilization, 2008
- Initiation of supporting regional FAO project on irrigation modernization, 2008
- Initiation of supporting FAO project on river basin allocation financed by MAFF/ Japan, 2008

How to Assess the Performance/Progress

- Progress of action plans in basin and functional centers of excellence and reference
- Meetings, workshops, raising of external resources
- Assess improvement in social, economic and environmental outcomes of irrigation at different levels according to guidelines for monitoring investment and results developed under the initiative and applied at national, basin and local levels
- Capacity building indicators (number of people trained, workshops, etc.)

How will the Initiative be Funded and Sustained

Hard pipeline and confirmed funds are already available for the following:

- Initiation of supporting regional FAO project on irrigation modernization, in 2008 (US$450,000)
- Initiation of supporting FAO project on river basin allocation financed by MAFF/ Japan, mid-2008 (US$ 900,000)
- Core funding from FAO's Regular programme for developing and initial secretariat/coordination services

Additional funding will to be sought for:

- core activities related to initiative management and networking activities at various functional levels by FAO-IWMI
- documentation, cases studies, capacity building, injection of knowledge synthesis and research results, issue-oriented research and development of capacity building capacity for basin and functional centres of excellence and reference
- technical activities related to topics not yet covered
- networking of regional and international knowledge centres (expected from KRA1 knowledge hub initiative)

Implementation/application activities at national, river basin, local government, and functional centres of excellence are intended to be financed out of available local resources. The initiative intends to support and improve on-going plans, programmes and activities by functional centres of excellence, rather than raise financing for implementation. The initiative will also develop the basin and functional centres of excellence as attractive locations and partners for relevant initiatives aiming at operationalising research and capacity building. Capacity building by networked international and regional knowledge centres outside the basin and functional centres of excellence and reference will be financed by training recipients. The initiative will essentially direct clients seeking capacity building to relevant knowledge centres.

Through this funding strategy, continuous funding for initiative management will be kept to a minimum, and external financing will be focused on supporting activities in the centres of excellence and reference.

Concrete Initiative 2 - launched by IUCN

Initiative Title

A Community of Practice in Asia-Pacific: Environmental Flows for Water Resources Management

Specific Objectives

Over the last five years, IUCN has been working to promote and implement environmental flows. Environmental flows means enough water is left in the rivers in order to ensure downstream environmental, economic and social benefits. In September 2007 at the River Symposium in Brisbane, Australia, IUCN and our partners launched a Global Environmental Flows Network. This network connects water managers, governments, businesses, NGOs, local communities and scientists to share knowledge or experience on environmental flows. The recently launched website (www.eflownet.org) will provide knowledge of environmental flows to new audiences by offering readily available access to current practices and methodologies.

The key objective of the IUCN Initiative is to form a community of practice in the Asia Pacific region on environmental flows for practitioners and institutions, to provide them with the power and capacity to lead change and innovate water management practices. The Asia Pacific experience from the Initiative will aim to inform the Global Environmental Flows Network.

Implementing Organizations

IUCN has been working with a number of partners to promote environmental flows globally and regionally. IUCN will continue to work with these and new partners including NARBO, IWMI, WWF, FAO, The Nature Conservancy, UNESCAP, the Mekong River Commission.

Active/Potential Partners at Sub-Regional Level/Country/Sub-Country

See above

Implementing Schedule Before and After the 1st APWS

- Introduce concept and solicit feedback at working session in Bangkok on 26 September
- Share idea and garner feedback at the APWS workshop on 27 September, Bangkok
- Firming up of concept note and partnerships, now until 15 October
- Preliminary meetings at 3rd Southeast Asia Water Forum, Kuala Lumpur, 21–25 October (tentative)
- Link into activities of IUCN Environmental Flows Network
- Incorporate into proposal for 2nd phase of IUCN Water and Nature Initiative

How to Assess the Performance/Progress

- Development of indicators of success (# of people from region formally joining the knowledge hub, integration of AP issues into global network, annual questionnaire, # of requests for information)
- Development of successful case-studies where e-flows has been adopted in legislation or in project development (eg. Incorporation into national water policies, World Bank/ADB guidelines or private sector guidelines)
- Increase funding opportunities for environmental flows assessments, policy change or capacity-building activities
- Increased multi-stakeholder dialogues which address e-flows at various levels (eg. local, national, regional)
- Expansion of partnership base for initiative

How will the Initiative be Funded and Sustained

- Incorporation into global Environmental Flows Network which was recently launched at the RiverSymposium in Brisbane, Australia
- Incorporated into funding proposal for IUCN Water and Nature Initiative Phase 2 (submission in mid-October for an 80 million Euro global program)
- Incorporation into IUCN Dialogues Program with Finnish government (1 million euros over 3 years)
- Discussion for incorporation into MRC IBFM program and linkages
- Translation of FLOW toolkits already commenced in Mekong Region and funding will be sought for other languages via WCD prize funds and others

The key activities of the Initiative are:

- To provide resources to find and share information on environmental flows and how it applies to the Asia Pacific region;
- To provide access to environmental flows related tools;
- To support stakeholders in developing solutions to water allocation leading to greater water security and equity between users;
- To create space for dialogues among key decision-makers an innovators;
- Continue to translate relevant materials into relevant languages;
- To build capacity and provide training, using toolkits for decision-making;
- To adapt and use best practices from the region to achieve integrated water resources management (IWRM).

Leadership for Ensuring Water Security in the Asia-Pacific Region: Knowledge, Financing and Capacity Development: Priority Theme A and Key Result Area 1

organized by Asian Development Bank (ADB), PUB Singapore, United Nations Educational, Scientific and Cultural Organization (UNESCO), Japan Bank for International Cooperation (JBIC) and Japan International Cooperation Agency (JICA)

Objectives

- To rally for leadership and commitment, show case success stories and institutional frameworks for financing water related infrastructure and human capacity development.
- To bring commitment by leaders on concrete programs to also invest in capacity development for people and institutions that build self-reliance.
- To facilitate dialogue between leaders from the region, and partner organizations to tap new sources and to develop innovative financial mechanisms.
- To bring commitment by leaders to facilitate development of knowledge networking and knowledge hubs.

Agenda

Curtain Raiser Film/Introduction by Moderator

Welcome Remarks by Mr. Koji Tanami, Governor of JBIC

Opening Remarks by Mr. Khoo Teng Chye, Chief Executive of PUB Singapore

Keynote Speech by Mr. Haruhiko Kuroda, President of ADB

Roundtable Panel Discussion

MODERATOR: Prof. Maria Mercedes Robles

PANELISTS:

- His Royal Highness Prince Willem-Alexander of The Netherlands, Chairman of UNSGAB
- Hon. Mahendra Pal Chaudhry, Minister of Finance, National Planning & Public Enterprises & Sugar Industry, Republic of the Fiji Islands

- Hon. Tetsuzo Fuyushiba, Minister of Land, Infrastructure and Transport, Japan
- Hon. Mahinda Dharmasiri Bandusena, Secretary to the Prime Minister, Democratic Socialist Republic of Sri Lanka
- Mr. Ek Sonn Chan, General Director, Phnom Penh Water Supply Authority, Kingdom of Cambodia
- Ms. Margaret Catley-Carlson, Chair, GWP

Leaders Commitment

- H.E. Ms. Khempheng Pholsena, Minister to the Prime Minister's Office, Head, Water Resources and Environment Administration and Chairperson, Lao National Mekong Committee, Lao People's Democratic Republic
- Hon. Dr. Ram Sharan Mahat, Minister of Finance, Nepal

Closing Remarks

- Mr. Richard Meganck, Director of UNESCO-IHE, Institute for Water Education
- Mr. Ariyuki Matsumoto, Vice President of JICA

Summary of Discussions

After a warm welcome for His Imperial Highness Crown Prince of Japan and His Royal Highness Crown Prince of Orange, the session opened with a curtain raiser film about key messages on water security in the Asia-Pacific region.

Opening Speeches

Welcome Remarks. JBIC Governor Koji Tanami shared that Japan is one of the largest donors extending assistance in the water sector in Asia wherein about 20% of JBIC's total ODA loans is for the water sector. It is estimated that $8 billion-investment annually would be needed in Asia to achieve the MDGs. Doing so is not impossible. However, to solve water problems, financing is not the only requirement. Each country has to have strong political commitment to set its own policy and institutional framework as well as to promote transfer of water resource management technology and know-how. In this respect, ADB, Singapore PUB, and UNESCO-IHE have great roles to play. He informed that JBIC and JICA will be merged in October 2008 to deal with all modalities of assistance, namely, ODA loans, grant assistance, and technical assistance, allowing the new agency to make even larger contribution to the water sector.

Opening Remarks. PUB Chief Executive, Mr. Khoo Teng Chye shared the success story of Singapore water management which is attributed to its holistic approach to water management. He also reported the progress of KRA 1. PUB has conducted capacity building workshops to share and exchange knowledge on water resources management with more than 520 participants from 45 countries since the establishment of the APWF in September 2006. He mentioned that one of PUB's key commitments as a lead agency for KRA 1 is to set up a network for knowledge hubs. PUB reported that their Water Hub - center for learning, technology, and networking has been established as the first knowledge hubs on urban water management. The next step is to establish other regional hubs. He suggested that the next milestone for KRA 1 could be at the inaugural Singapore International Water Week which aims to be an annual global platform for water solutions. The theme for next year's Singapore international Water Week which will be held in June 2008 is "Sustainable Water Solutions for Cities". He extended a warm invitation to all to be at the inaugural Singapore International Water Week.

Keynote Speech. ADB President Haruhiko Kuroda called on all countries in the Asia and Pacific region to make a fundamental change in the way they manage water to build a sustainable future. ADB regards better water management as a crucial challenge for the Asia and Pacific region. Calling for strong political will and practical, forward-looking actions, Mr. Kuroda stressed the need to move water higher up on the local, national, regional and international policy agendas. He also mentioned that ADB is committed to help its developing member countries address their water challenges with the vision of "water for all" in the Asia and Pacific region. And as part of ADB's contribution to the Summit, ADB released the Asia Water Development Outlook on November 29 – a report that assesses the region's current and future water problems and proposes the policy measures that can help solve them.

Panel Discussion

The 45-minute panel discussion collected the views and opinions from His Royal Highness, Prince of Orange and decision-makers and water champions from the Asia-Pacific region on the different issues/messages relating to achieving water security in the region. Based on the responses, experiences and knowledge shared by the panelists the following key messages were given importance in achieving water security in the region:

a. *On the importance of water governance in addressing the water challenges in the Asia-Pacific region.* The old ways of engineers alone are not going to fix the problems. We need the engineers, the donors and the political framework to achieve the MDG. We need the software, the brainpower behind the hardware, to achieve the MDG. (*His Royal Highness Prince of Orange*)

b. *On the importance of water resources management.* Effective use of existing facilities is quite important tool. For instance, through installing new gate structure, the unutilized water area, so-called "dead water area", is diverted to other effective use for flood control or, where water is scarce, precious water resource. As another example, removal of sediment in dam reservoir by using the by-pass secures the available capacity of dam reservoir. In addition, we have started the efficient reservoir operation by effectively reallocating storage capacity of individual dams that are strong in flood control function or water supply function, according to the regional rainfall pattern. Japan is using this engineering technique in order to prevent flood and make best use of invaluable water resources. (*Minister Fuyushiba*)

c. *On the importance of knowledge and database.* The potential users of water have to be integrated. In other words, you just cannot let the agriculturists doing what they need to do and the energy people doing what they need to do without regards to the floods and to the transportation and the other dynamics. We need to put these together and we cannot put them together unless you have database as knowledge. The important part in the 21st century is not necessarily to create this but to find out where they are and how to use them and that is the important part of what Mr. Kuroda and Mr. Khoo were talking about - to create a knowledge hubs. You It should be ensured that those hubs are informed, they get the knowledge they need, and help people understand how to go and use that knowledge and data. It is creating expertise and comfort zones in the users of this knowledge so that they would go in partnership with others to go and use the data and the knowledge that has been created. So the complete phrase is "integrated water resources management". It depends on getting people together and on using good data and information. (*Ms. Catley-Carlson*)

d. *On how island nations are looking on supposedly a new paradigm in the solution of water problems such as IWRM and new forms of governance.* Water governance is assuming priority in government national agenda as we deal with climate change as you know many of us, island states are threatened with submergence on account of rising water level. In 2003, the region endorsed a plan for water governance known as South Pacific Regional Action Plan for Sustainable Water Management. Each of the island nations have input to the plan and it was unanimously endorsed by the island leaders. And we are now working on that plan and you know to improve governance, of course we have constraints because what you can do largely depends on what resources you have at your disposal. Island nations are starving of resources in all sectors, in terms of knowledge, governance issues and in terms of financing. Soft loan consideration from financial institutions is being appealed for small island nations as affordability for infrastructure development by way of loan finance is very limited. There are a lot of competing demands on government finances which comes from the social sector, education, health, and welfare. Financing is very crucial to island nations. Island nations will depend on institutions like the international water forum, and those with necessary skills in this area to help them along the other technical areas but the most important issue for the island nations is of course financing. (*Min. Chaudry, Fiji*)

e. *On how important the water issue for Sri Lanka and what efforts were done to address these problems.* Sri Lanka is an island nation which use ample amount of rainfall but unable to harness water resources at most benefit to the entire nation. There is a certain amount of antipathy for the private sector to come in investment in water resources and sanitation. The bulk of investment has got to be undertaken by the government as it has been the tradition for the government to undertake investment. The National Water Supply Drainage Board (NWRDB) undertakes the heavy investments in providing large scale water services in drainage and water supply and sanitation services. International donor agencies, especially JBIC, JICA, and ADB, has extended their assistance in extending soft loans to NWRDB, undertaking 15 big projects that are being implemented all over the country. Besides the central government organizations, the local authorities, and provincial government organizations undertake minor irrigation water service activities and sanitation activities and that is not sufficient. Further investment is required. Specially, the private sector should come in to the picture and should be in a position to invest to small projects specially, with the tsunami coming into wherein Sri Lanka is one country that has been affected. Donor agencies and NGOs came forward to bring help to the people in the coastal areas and they undertook various water and sanitation projects and that has now given some indication that not only the government but also the private sector and the NGOs would bring certain amount of inputs to increase the level of water service and sanitation in country. There is really a need for assistance and new partnerships to be forged. (*Sec. Bandusena, Sri Lanka*)

f. *On what is the formula behind the turnaround for Phnom Penh Water Supply Authority (PPWSA).* From a civil war of 20 years, around 700,000 people in 1979 people came back to Phnom Penh without water. The situation of water supply in 1993 was really very bad. The capacity was only 63,000 to supply to 700,000 people. The network is 8,000 km all cast on pipe bring around 72% of NRW. The quality of water was terribly bad. From then, the government set a new policy market-oriented set up of policy for water. Financial and other assistance from other donor namely ADB, JICA and as well as some government came in.

Especially, in the recent years, JICA provided a capacity building programs which brought PPWSA's management capacity up. Through political will, financial assistance from donors, and the improvement of management capacity practices of how to operate it, PPWSA is now in a very sustainable way. PPWSA can produce enough water for about 1.5 million people for Phnom Penh as of today. Water loss is only around 7% with a collection ratio of almost 100%. (*Mr. Ek Sonn Chan*)

g. *On the specific measures that will address the need for financing mechanisms.* In general, one of the fields of UNSGAB, one of the 6-pt Hashimoto Action Plan is financing. Although ADB is very positive and has been putting a lot of money into it. Also, through water operator partnership, private sector is very willing to put money into water and sanitation project, however, if we need to get the right regulation and the guarantees from the government. And from regional development banks, we really have to look at new ways of financing and try to get it at local financial market because we're trying to find ways but that is the most difficult aspect of water management. (*His Royal Highness Prince of Orange*)
h. On the concerns from developing countries that financing will not come if they have not raised/developed their capacities.
i. One of the ways of financing will be with donors support, and the only way to build capacity is with donors support. The market will not function there in the first place. Having come said that, there is always this aspects of should water be free? Water is free when you collect it yourself but if it is serviced to your house through pipes the servicing cost has to be paid. This feeling that water is free is something that came from the 2nd half of the 20th century when politicians used water as a political tool promising free water and actually running down the public utilities in the meantime not giving them a chance to create revenue, to cost recovery for operation and maintenance so the system run down and then at the end of the day, the poorest of the poor, the slums, are paying 10 times as much on dubious source out of water tanks because we mismanage the system. The people in the slums are paying more for water now much more than they should do when they have piped water. Every sustainable water connection is a good one. It doesn't matter how it is financed as long as it is sustainable and affordable, there should be cost recovery for the water operator and subsidizing system (for example in South Africa the 1st 25 liters are free), but at the end of the day the water providers also needs a way to come out with its cost. (*His Royal Highness Prince of Orange*)
j. *On balancing financing and capacity development.* Underlying the what Mr. Ek Sonn Chan was said, what are the ingredients in his success factor, Phnom Penh was one of the poorest, least resourced cities in the region as he said terrific

challenges coming out of an extended civil war period. What are the ingredients? There were some international assistance to infrastructure, but basically it was knowledge, support and he and his people were the ones that got up to 100% billing and down to 7% of leakage. It is very possible to invest in water operators and use the examples of Phnom Penh and Singapore to extend to water operator the kind of capacity that they need to get up to the point where they can start to go on local capital markets. There is also private money available, don't just think about the private water operators, think about pension plan, think about national capital sources, about local capital markets. Operator has to have both internal capability of improving their operations and the ministers of planning and finance have to look at the conditions in which they operate to make sure that decentralization to them is real. In this region there are several institutions ideally to take on accelerated program of helping utilities. So it's no miracle on how to get finance to flow in. The capability is there, the good part should be focused on improving water operators so that they can emulate these good examples. (*Ms. Catley-Carlson*)

k. *On the possible obstacles in Sri Lanka in replicating the success stories of Cambodia and Singapore.* There is a certain misconception from the people that interaction with the private sector in this are will jack-up the price they charge in water so the people are more inclined to investment from the public sector. Hence, the local authorities should be empowered so that they can undertake water projects with community participation and thereby improving the water situation in the rural areas. In urban areas, the NWSDB can undertake the big projects. (*Sec. Bandusena, Sri Lanka*)

l. *On the possible obstacles in Fiji in replicating the success stories of Cambodia and Singapore.* Fiji is a poor island and it is very difficult to attract private investments on the account of lack of volume. There is also tremendous public pressure for water to remain in the hand of the state. But the important thing is if we have efficient water system and affordable water rates, people don't mind paying for water. We have abundant water, but before going to big investments, we should look first our ecosystem. If we able to preserve, maintain, and use the resources we have in our ecosystem, then that would be the most effective way of dealing with it. What Fiji is doing at the moment is to see how nature can help in solving the problem. They need expertise to understand how the ecosystem works. This is where capacity building is needed in getting water closer to the people.

m. *First comment from the audience.* The workforce in public water utilities are among the best sources of knowledge on water. They are the ones who run the water operations, they interact very closely with the communities that they serve, they have the personal knowledge on how to improve the water services. Workers and their union should be adequately represented in all water partnerships and knowledge hubs and public partnerships should also be promoted on a not for profit basis.

Response from His Royal Highness. It's exactly UNSGAB is working on in its water operation partnership, because in the past, the workers and the trade unions have not been heard. We have to try different approach to supplying water to public utilities and if we do not use the knowledge of the workers and the trade union can come along with that, but its really the workers that should really work on that and if now we launch the global water operator partnerships initiative under UN-HABITAT and we always monitor from our UNSGAB board that the workers will have their place they deserve in this place.

n. *Second comment from the audience.* Another comment. Water for all, the biggest challenge would be to provide water access to the poor. How can this goal be addressed? How can these initiatives be supported?

o. *Response from Min. Fuyushiba.* The population and industries are concentrated in the Yodogawa River Basin in the Kinki region. How to supply and manage water is a big issue. The intake for drinking water and drainage of sewage disposal are repeated along the river, and approximately half of 11.8 million habitants in the basin drink water which is recycles six(6) times. Nevertheless, an excellent sewage disposal technology and water treatment technology enables the residents to drink a safe, good taste water, directly from the taps. The tap water of Osaka in the downstream is sold as a commodity, with the PET bottled. In Japan, with development water supply system, patients of Cholera, Dysentery and Typhoid fever and the death rate of infants have decreased. Japan is willing to support the Asia-Pacific region, which has similar climate and culture, offering the know-how and the technology.

In wrapping up the panel discussion, a question was addressed to the three leaders in the panel: "What messages are they going to take with them from the summit to share with their leaders back home that will hopefully have some impact on the way their country are facing boldly and bravely the problems of water?"

p. *Response from Mr. Ek Sonn Chan.* "Phnom Penh tried to help ourselves along the line with the help we get from different donors. The government of Cambodia knows how to help ourselves. Everything could be happened in the right direction if we can control ourselves and we have the mind to love our people."

q. *Response from Sec. Bandusena.* "While following this session, I learned the importance that is attached to water. And I would say that I would take back and impress on our leaders the support and knowledge that we can get from Japan and organizations like PUB of Singapore, practical experiences that our colleague have through the years so that the leadership in our country get lessons out of these experiences and so that they would themselves giving more access to water to greater part of the population of the country. Its our aim actually by 2015 that we would be able to achieve the MDG but I think we have to work in an accelerated form as far as possible the government will have the cooperation of the community and the private sector to work in the horizon to improve water resources projects so that both the turban and rural sector will get the benefit of accessible water."

r. *Response from Min. Chaudry.* "This summit is an eye opener for me. I must admit. And I must thank the organizers and sponsors of the summit. We held world water summit but we don't have one for the region. I am not an expert in this field but now I know and I am informed a little better. And I am sure that what I gained here will go a long way in helping us in addressing the problem that we face and I am sure that the regular forum of this nature will assist all of us in overcoming our water worries."

The session moderator, Prof. Maria Mercedes Robles, concluded the panel discussion by saying that we all hope that the summit conveners and the organizers as well as all the water experts and water advocates and some of our guests here on stage are leaders of those initiatives. May the efforts service beacon to illumine the way for many countries that really need a lot of assistance in addressing the water problems that beset the world's most diverse, most dynamic, and fastest growing region - the Asia-Pacific.

Leader's Commitment

Her Excellency Madam Khempheng Pholsena, Minister to the Prime Minister's Office of the Lao People's Democratic Republic, expressed her country's commitment to ensuring the sustainable development of Lao PDR's water resources in order to contribute not only to improved health and sanitation of our citizens and generate income to invest in poverty reduction in their country, but also to contribute to the Asia-Pacific Region's water security and development through strengthened regional cooperation at all levels with related stakeholders within the framework of the Asia Pacific Water Forum, ASEAN, Global Water Partnership, Mekong River Commission, GMS and other forums as well as bilateral and multilateral cooperation.

Honorable Nepal Finance Minister Dr. Ram Sharan Mahat opined that leadership plays a crucial role to ensure water security in the Asia and Pacific Region. The role of leadership is evident at various levels: local, national, regional, and global. He cited an example of how leadership at the local and national level has brought about transformation in the landscape of mountain region of Nepal. He expressed support to the concept of knowledge hub. He also mentioned that Nepal has hosted the Headquarters of International Centre for Integrated Mountain Development (ICIMOD) which is doing a good work in this field, and Nepal will continue to support them. Finally, in general, Nepal would support the holistic approach to water management: basin management, intra-regional, optimization of basin water, capping of monsoon flows and raising dry season flows, multiple use, and energy generation.

Closing Remarks

Director Meganck mentioned that we are at a critical juncture in our debate on the overall topic of leadership – whether we are speaking of water security or any other water-related topic and leadership. If we are truly interested in the long-term, in the sustainability of such investments, then leaders of all levels are required and in sufficient numbers. The need to develop the knowledge hubs is vital in the regions and countries where the problems are being encountered. Additionally, we need to prioritize local actors and build on local knowledge to encourage water leaders and champions. We also need to raise awareness through teaching water literacy. We must speak more forcefully about interrelatedness and dialogue among water stakeholders. And finally, we must share information and knowledge. Whether we like it or understand it, we are living in an age where data, all data is available. Someplace all of our shoe sizes are recorded. Open access software, journals and technical information is the future. Let's embrace it.

Vice President Matsumoto closed the session by stressing that water problem is closely associated with the achievement of all aspects of MDGs, not only the issue of water and sanitation issues but also issues of health, education and poverty reduction. Improved access to water significantly contributes to the realization of human security. Capacity development, including establishment of a fair policy framework, forms the foundation for the appropriate operation and maintenance of water supply facilities and encourages sustainable investment for water infrastructure. Finally, he attested that JICA will strive its utmost to provide result-oriented and speedy cooperation in order to mitigate water problems, achieve MDGs and realize human security.

Concrete Initiatives

Initiative Title

1. Financing Water and Sanitation and strengthening capacity development
2. Establishment of initial knowledge hubs on water

Specific Objectives

For Initiative 1

1. To reach 200 million people with access to sustainable safe drinking water and sanitation through ADB's Water Financing Program 2006–2010.
2. To support the self-help efforts for the capacity development on water and sanitation under "Water and Sanitation Broad Partnership Initiative (WASABI)" launched by the Government of Japan in 2006.

For Initiative 2

1. To develop a network of centers of excellence (Knowledge Hubs) for knowledge networking in the Asia Pacific region.

Implementing Organizations

For Initiative 1: ADB, JBIC, JICA
For Initiative 2: Singapore PUB, UNESCO

Active/Potential Partners at Sub-Regional Level/Country/Sub-Country

For Initiative 1

UN-HABITAT (under ADB-UN-HABITAT partnerships); GWP-SA, SEA, NA Pacific, and CA (for advocacy and capacity development); local institutions in key developing countries (to be decided), UNESCAP (for coordination with IYS 2008, and monitoring of investments and results under APWF's KRA4), and development partners (Netherlands, Australia, Spain, UK).

For Initiative 2

International Centre for Water Hazard and Risk Management (ICHARM, Japan), International Research and Training Centre on Erosion and Sedimentation (IRTCES, China), The Energy and Resources Institute (India), Yellow River Conservancy Commission (China), International Water Management Institute (Sri Lanka), K-Water (Korea), Mekong River Commission Secretariat (Lao PDR), Ministry of Public Works (Indonesia), Ministry of Natural Resources and Environment (Malaysia), Scientific Information Center of the Interstate Commission for Water Coordination of Central Asia (SIC-ICWC, Uzbekistan), South Pacific Applied Geoscience Commission (SOPAC, Fiji Islands), Japan Water Agency (JWA), Asian Development Bank (ADB) and UNESCO-IHE.

Implementing Schedule Before and After the 1st APWS

For Initiative 1

- Implementation of ADB Water Financing Program to be open with potential new partners.
- Annual consultations between three institutions and other development partners.
- The outcome of APWS will be used as an input for the discussion of the relevant issues at the 2008 G8 Summit in Toyako.
- JBIC and JICA will utilize the outcome of the APWS in their future dialogues on the implementation policy with the governments of developing countries to realize effective formulation and implementation of projects.

For Initiative 2

- Consultation meeting with some 12 candidate knowledge hubs was held from 29–31 Oct 2007 at PUB Singapore's Waterhub. The participants of the meeting expressed interest in collaboration among the candidate water knowledge hubs and recommended the establishment of an APWF Network of Regional Water Knowledge Hubs.

How to Assess the Performance/Progress

For Initiative 1

- A Result framework will be developed for the initiative and progress will be reported in the future Asia-Pacific Water Summits.

For Initiative 2

- Knowledge Hubs will be established and progress will be reported in the future Asia-Pacific Water Summits.

How will the Initiative be Funded and Sustained

- Under the ADB's Water Financing Program 2006–2010, ADB intends to invest about $10 billion in the water sector, supported by $100 million in grants from the Water Financing Partnership Facility.
- Each knowledge hub secures its own financial arrangement. (For 2.)
- In October 2008, JICA and JBIC will be integrated as "new JICA" which will comprehensively deal with Japan's ODA loan, grant aid, and technical assistance. In view of that, both JICA and JBIC are currently jointly formulating country-specific strategies so that different types of assistance will make a synergistic effect in recipient countries.

Increasing Capacity for Local Actions: Key Result Area 2

organized by United Nations Human Settlements Programme (UN-HABITAT) and Streams of Knowledge (SOK)

Objective

The key objectives of the Session were:

- To drive home the main messages of KRA-II
- To generate better understanding and support for the KRA-II messages
- To launch the capacity building hubs initiative for future exchange of information and knowledge and capacity building for implementation.

Agenda

The major thrust of KRA-II is to provide universal access to safe drinking water and basic sanitation to all by focusing on (i) strengthening decentralization with empowerment; (ii) promoting partnership between local authorities, civil society (NGOs and community-based groups) and different service providers (public and private); and (iii) capacity development as an important area of investment.

The Video Montage titled 'Bridging the Gap' focused on the challenge and the increasing the capacity of local authorities, utilities and communities.

In his welcome remarks, Mr. Andre Dzikus, explained the purpose of KRA II Session at the APWS and invited the group of distinguished panelists to discuss the issue of increasing capacity for local actions and suggest the approach and way forward for the concrete initiative.

The keynote address by Dr. Anna Kajumulo Tibaijuka, read out by Mr. Toshiyasu Noda, underscored the need for generating new ideas for developing and operationalising a capacity building strategy for implementation and expressed commitment of UN-HABITAT to team up with other organizations and institutions to establish national and regional capacity building hubs in order to develop knowledge products and also assess the progress of capacity development interventions in the region.

The Session Round-table Panel discussion was moderated by Ms. Mari Christine, UN-HABITAT Goodwill Ambassador in Japan. Prior to the panel discussion, there were three presentations on the following themes:

- Decentralization with Empowerment by Mr. Lito Atienza, Secretary of Environment and Natural Resources
- Partnership approach between Local Authorities, Civil Society and different service providers by Dr. S.A. Hashim, President, India Water Partnership
- Capacity Development and Capacity development hubs as an important area of investment by Dr. Rivka Kfir, Streams of Knowledge

PANELISTS:

- H.E. Mr. Saif-ud-Din Soz, Minister, Water Resources, Govt. of India
- Dr. Somphone Dethoudom, Director General, MCTPC, Govt. of Lao PDR
- Mr. Zenkichi Kojima, mayor of Shizuoka City, Japan
- Dr. Sara Ahmed, Chairperson, Gender Water Alliance
- Mr. Henry Northover, Head of Policy, Water Aid
- Professor Tetsuro Tsujimoto, Department of Civil Engineering, Nagoya University
- Professor Komatsu, Kyushu University
- Mr. Andre Dzikus, Chief, WSS II, UN-HABITAT Nairobi
- Ms. Rory Villaluna, Executive Secretary, Streams of Knowledge, Manila

Summary of Discussions

In many Asian countries, the delivery of basic services such as water and sanitation has been delegated to local authorities. However, there was still a big gap between the decentralized responsibility and the current capacity to support their role. There was a felt need that national governments should focus on strengthening local level delivery systems. Local authorities must be equipped with an enabling environment to ensure sustainable service provision for all. The political will to assume responsibility needs to be harnessed and their capacities also need to be strengthened to ensure effective, sustainable and efficient service delivery systems and good governance at the local level.

Local authorities have been generally expected to ensure basic services. The rights-based approach demands that the state is expected to promote, respect and fulfill this right. However, it is not expected that they provide the service themselves. The key is to open up to multi-stakeholder participation and community-led initiatives. Collaboration is crucial at different levels to achieve mutually accepted goals of sustainable service delivery for all, especially targeting the poor and marginalized groups of men and women, children, the sick and the elderly. Participation in planning, budgeting, technology choices, performance goal setting and monitoring are some of the key areas that need to be opened up for collaborative action. Partnership approach among local authorities, civil society (NGOs and community-based groups) and different service providers (public and private) need to be encouraged. The major highlights of the discussions of the KRA-II Session are briefly given below:

The focus of the session remained on capacity development of local authorities, utilities and communities with the purpose of providing people with knowledge to assist them in making informed choices through development of appropriate technology, training modules, resources databases, conducting action research and making policy recommendations. There was an emphasis on functional and financial decentralization for empowering local authorities and strengthening the role of civil society self-help groups, resident welfare associations and communities at large. One of the key recommendations was the partnership approach between local authorities, civil society and various service providers. In pursuing various goals for strengthening local capacities, the involvement of women with a gender balanced approach was considered quite necessary, besides adopting a holistic approach to meet the needs of various sectors. Fiscal decentralization which may help local control on resources for water and sanitation and also build capacity for planning, budgeting and project management. There was an emphasis on mobilizing political will to improve accountability and aligning international, national and local efforts and also build sector wide transformation. National initiatives in countries like Lao PDR are supporting community based schemes and promoting private sector participation.

The panelists took note of the fact that sustainable water and sanitation was not just about infrastructure. Most Asian governments so far have failed to provide resources for capacity development to ensure that the knowledge, skills and behaviour changes were in place. There was a recognition that adequate investments in capacity development shall contribute to the ability to work with informed partners, work at scale, replicate good practices and innovate. It will also enable the local actors to provide the necessary support system to follow-through and ensure continuity of an environmentally sustainable service delivery system.

Messages

The major outcome of the KRA-II Session was the need to enhance learning and to replicate and scale up local actions for sustainable improvement in the water and sanitation sector for enhancing efficiency of the utilities, local authorities as well as the communities to reach the last mile with the purpose of increasing the coverage, with special emphasis on the poor. The main messages of the session were:

- Strengthening decentralization with empowerment;
- Promoting partnership between local authorities, civil society and different service providers; and
- Capacity development of local authorities, utilities and communities.

Concrete Initiative

Initiative Title

Capacity Building Hubs and Policy Support for Increased Local Capacities

Specific Objectives

The concrete initiative relating to Capacity Building Hubs for Increased Local Capacities cuts-across all the three Themes of Water Financing; Water and Disasters and Water Development and Ecosystems. The Capacity Development hubs will serve as an active link with the various country level NGOs, civil society and local governments to support the advocacies of the three themes and may also be mobilized to support KRA 4 on monitoring investments as well as KRA 3 for increasing public outreach.

The specific objectives of this Concrete Initiative are (a) sharing knowledge about the local actions/best practices in water and sanitation which are successfully implemented in various countries of the region; (b) to support capacity building hubs at country and local levels to facilitate future exchange of information and knowledge and sustainable service delivery and capacity building for implementation; and (c) support policy makers in promoting effective local action through pilot and demonstration projects.

Implementing Organizations

The main implementing organizations of this concrete initiative are UN-HABITAT in collaboration with Streams of Knowledge.

Active/Potential Partners at Sub-Regional Level/Country/Sub-Country

Asian Development Bank, Asia-Pacific Water Forum, World Water Forum and other partner organizations are the likely partners at the regional, sub regional, country and local levels.

Implementing Schedule Before and After the 1st APWS

Before 1st APWS

UN-HABITAT and Streams of Knowledge organized one day workshop with the lead organizations and Secretariat for finalization of programme in Bangkok. Additionally, a workshop for the water champions was organised on 23rd October 2007 during working group meeting for finalization of the APWF Policy Brief 2007 in Kuala Lumpur from 21–27 October 2007.

After 1st APWS

Based on the recommendations of the KRA-II Session, a detailed plan of action has been drawn up for implementing various activities for Increasing Local Capacity. The action plan, inter alia, includes the following:

(i) Identify and work with institutions in the region in building their capacity to support local actions. These Capacity Development hubs will build on existing capacity development networks at various levels to support KRA-II's plans that include raising awareness amongst local authorities, water and sanitations service providers, civil society and communities. The implementing organizations together with APWF shall invite them to support the process of mapping networks and associations of local authorities; identifying local champions; conducting a rapid capacity needs assessment of local authorities, water and sanitation service providers and community/women groups/civil society; identifying/supporting other potential Capacity Development Hubs at regional, sub-regional, country and local levels.

(ii) As a part of this initiative, an open access web-site and newsletter for "effective local action" for knowledge sharing shall be developed.

(iii) A manual for effective decentralization and empowerment for use by policy makers is proposed to be developed.

(iv) Reinforce effective local action as a theme in UN-HABITAT supported pilot and demonstration projects.

Identification of institutions and organizations at the regional, sub-regional, national and local levels which will serve as the capacity building hubs for sharing knowledge about the local actions/best practices in water and sanitation. These hubs will facilitate future exchange of information and knowledge and sustainable service delivery and also build capacity for implementation. The implementing organizations will support building the capacity of these identified capacity building hubs for effective delivery and will strengthen them to produce training manuals for meeting various capacity building requirements.

How to Assess the Performance/Progress

The main criteria to assess the performance/progress will be the increased efficiency of the water utilities and greater coverage in terms of accessibility of water and sanitation services.

How will the Initiative be Funded and Sustained

The initiative is proposed to be funded and sustained through National Governments with the support of bilateral and multilateral agencies.

1st Asia-Pacific Water Summit

Policy Brief 2007

Beppu, Japan, 3-4th December 2007

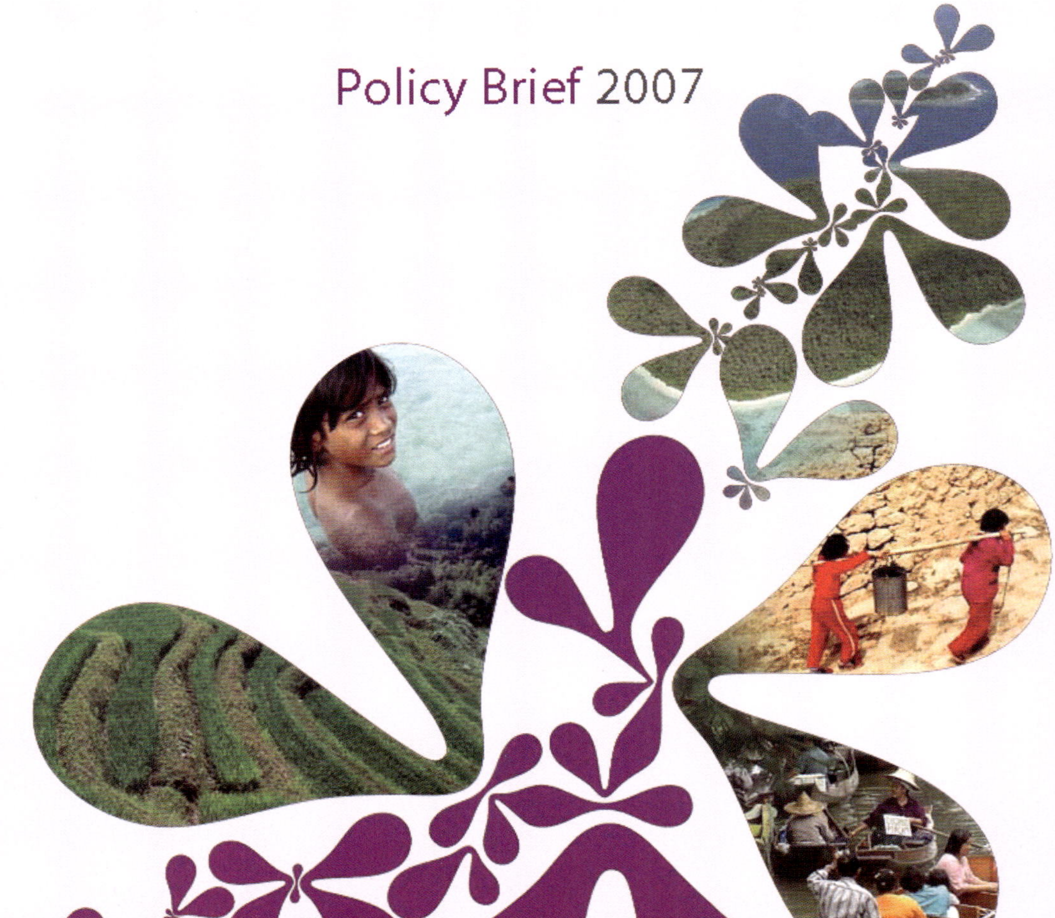
Policy Brief 2007

Main Contributors

Asian Development Bank (ADB)
The International Center for Water Hazard and Risk Management (ICHARM)
The World Conservation Union (IUCN)
Food and Agriculture Organization of the United Nations (FAO)
Public Utilities Board Singapore (PUB Singapore)
United Nations Educational, Scientific and Cultural Organization (UNESCO)
Streams of Knowledge (SOK)
United Nations Human Settlements Programme (UN-HABITAT)
United Nations Economic and Social Commission for Asia and the Pacific (UNESCAP)
Japan Water Forum (JWF)
Japan Bank for International Cooperation (JBIC)
Japan International Cooperation Agency (JICA)
World Health Organization (WHO)

Special thanks to subregional coordinators and those who engaged in discussions through weblogs and workshops.

Printed by Clung Wicha Press, Thailand

ADB

ICHARM

IUCN
The World Conservation Union

FAO
FIAT PANIS

PUB
Water for All: Conserve, Value, Enjoy

UNESCO
United Nations
Educational, Scientific and
Cultural Organization

STREAMS OF KNOWLEDGE
Global Coalition of Water and Sanitation Resource Centres

UN-HABITAT

United Nations
ESCAP

Japan Water Forum
日本水フォーラム

JBIC
JAPAN BANK FOR INTERNATIONAL COOPERATION

JICA

Contents

© Asia-Pacific Water Forum

A Call to Action from the Asia-Pacific Water Forum to the 1st Asia-Pacific Water Summit

1. The 1st Asia-Pacific Water Summit (1st APWS) provides an unprecedented opportunity to recognize the central role that water plays for our quality of life and the wellbeing of future generations. The key messages and recommendations presented in this Policy Brief draw a clear and realistic picture of the needs presented throughout our region and of the actions required in order to achieve the level of progress deserved by all. To be sure, the challenges are great, but with will and committed political support, all of our goals can be met.

2. The 1st APWS is not an end in itself. It is, however, a critical milestone that is part of the ongoing regional and international efforts towards human and economic development. Indeed, the outcomes of the 1st APWS shall feed directly into other high profile international events such as the 2008 G8 Summit and the 5th World Water Forum. With 2008 earmarked as the International Year of Sanitation, the timing of the Summit is ideal for generating a momentum that will carry on for years and decades to come.

3. By addressing and engaging decision-makers at the highest possible levels, the 1st APWS seeks to generate the support and commitment required to meet our region's water challenges and to serve as an example of regional cooperation to the rest of the world.

B ackground

4. The Asia-Pacific Water Forum (APWF) is an independent, not-for-profit, non-partisan, non-political network that was launched in September 2006 for the purpose of bringing water to the forefront of the political agenda. The first phase the APWF's activity has focused on the three priority themes identified in the regional document for the 4th World Water Forum as the initial steps towards sustainable development and economic growth: water financing, water-related disaster management, and water for development and ecosystems.

5. The APWF has a matrix structure centered on three Priority Themes and five cross-cutting Key Result Areas (KRAs). A number of highly recognized institutions from the region have committed themselves to supporting the APWF as "lead organizations", responsible for identifying strategies and initiating solutions to achieve progress under each of the Priority Themes. Such initiatives have been identified through the KRAs and the "lead organizations" responsible for supporting them. Secretarial support for the APWF has been provided by the Japan Water Forum.

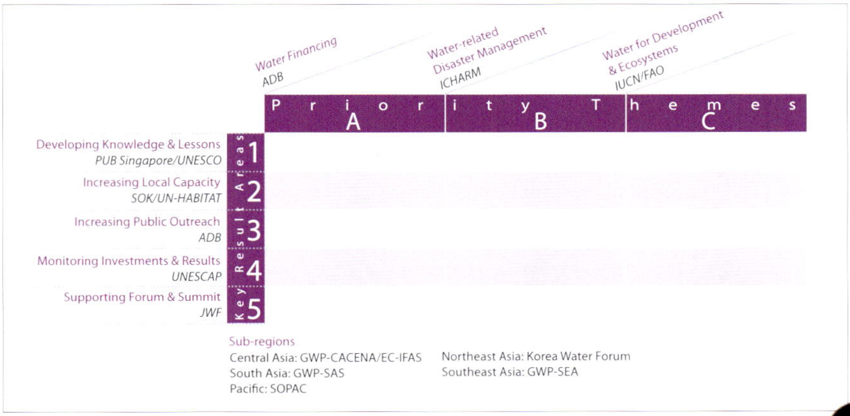

6. This Policy Brief outlines a series of recommendations prepared as part of the initial activities of the APWF in preparation for the 1st APWS, to be held in Beppu City, Japan, on December 3rd to 4th, 2007.

7. The recommendations herein set forth a clear path on how to meet the combined and complementary objectives of development and poverty reduction, economic growth and increasing prosperity, and environmental sustainability. These recommendations are expected to generate momentum in addressing the region's many water challenges and provide the basis for the adoption of a set of concrete initiatives that will not only represent a major output of the 1st APWS, but will make a true difference in solving our region's most urgent water problems.

Issues

8. Over the past few decades, several areas of the Asia-Pacific region have experienced notable progress in various aspects of water management, including access to water resources and coverage of related services. However, rapid population and economic growth, urbanization, and the underperformance of existing water assets mean that there are still significant shortfalls in meeting the region's needs. In addition, growth patterns and the development of water resources have severely affected both the environment and the region's ecosystems. The looming impacts of climate change will exacerbate these challenges unless they are properly taken into account throughout all planning and decision-making processes.

9. The Millennium Development Goals (MDGs) have set socioeconomic targets that need to be achieved by 2015. Although economic growth rates have been high overall, poverty remains a chronic problem in both urban and rural areas with 16% of the population in the region's developing countries undernourished. Progress towards halving the number of hungry by 2015 remains slow.

10. Access to improved drinking water supplies in the Asia-Pacific region has improved substantially in recent years (in 2002 around 3.16 billion people - 82% of the population - in the Asia-Pacific region had access to improved water supplies, up from 74% in 1990). However, formidable challenges remain. Approximately 669 million people live in the region, of which 584 million live in rural areas. Aggregate figures also mask dramatic disparities between the subregions, between nations within subregions, and within the nations themselves.

11. The story for sanitation is even less satisfactory. Of the 2.6 billion people in the world without access to improved sanitation facilities, nearly 2 billion are in the Asia-Pacific region. Coverage for improved sanitation in the region lags behind the rate needed to attain the MDG target. As of 2004, only 45% of the region's population had access to improved sanitation facilities – up from 43% in 2002. And of the 1,862 million people in the region that remain unserved, 1,498 million live in rural areas. These figures illustrate the need to develop new, innovative systems treating human waste that do not rely so heavily on the use of freshwater. And with 2008 being the International Year of Sanitation (IYS), there will be considerable momentum in the international community to provide assistance and guidance towards addressing this particular challenge.

12. The Asia-Pacific region is the most vulnerable region in the world with regards to water-related disasters that hinder sustainable development and poverty reduction. Between 1960 and 2006, over 600 thousand casualties were recorded, accounting for over 80% of casualties worldwide, in addition to US$ 8 billion worth of economic damage during the same period. Severe water-related disasters such as floods, droughts, tsunamis, windstorms, landslides, storm-surges, water-borne diseases, and epidemics have escalated since the turn of the 21st century.

13. Rapid population growth in the Asia-Pacific region over the past decade has forced more people to live in floodplains and other vulnerable areas and has led to ever-increasing demands for water supply, security and sanitation services, which require greater investment in water projects. In certain countries, investment in improved access to safe drinking water and decent sanitation facilities; allocation and conservation of water resources; adequate management of wastewater including its safe recycling; and construction of flood control and irrigation facilities, have led to increased economic growth; a healthier, more robust economy; and an overall reduction in poverty. Unfortunately, in most countries, investments in the water sector have fallen seriously behind demands.

14. One great challenge in moving forward as a region is the great diversity of its ecosystems, populations, political institutions, and socio-economic conditions. The region includes 38 major watersheds, of which 21 are transboundary. However, institutions concerned with shared basins are not capable of dealing with the political complexity of international water and the transboundary tensions created by large-scale water management schemes. At the other end of the spectrum, the isolated nature of small island developing states (SIDS) calls for increased regional cooperation to share knowledge and build capacity in order to address challenges common to many insular nations, as embodied in the Pacific Regional Action Plan on Sustainable Water Management.

15. Accounting for 84% of total freshwater withdrawals, agriculture dominates the sector. It is characterized by poor efficiency, low productivity, poor service provision, and non-existing or ineffective demand management. Many irrigation agencies are still driven by supply, and demand management mechanisms are either non-existent or ineffective. Furthermore, the intensification of agriculture, combined with industrial development and urbanization, have severely affected water quality.

16. Essential ecosystem services provided by aquatic ecosystems, upstream watersheds, and agricultural landscapes are being significantly degraded, especially in areas of high resource dependency, with severe consequences for the incidence and severity of water-related disasters, and affecting downstream urban and rural populations, and populations whose livelihoods depend on them. The aquatic biodiversity's high economic and social value is threatened by overexploitation and alterations in hydrological regimes.

17. Water resources are unevenly distributed, and water availability is complicated by physical, economic, and environmental water scarcity. Moreover, the region's freshwater resources – our river basins, watersheds, wetlands and ecosystems – are imperiled by pollution, inadequate management, and climate change. Indeed, it is highly likely that the additional stresses caused by changes in the region's meteorological and hydrological regimes will increase our vulnerability to water shortages and water-related disasters[2]. There is a need to make comprehensive and collective efforts to manage water-related disasters within the context of integrated water resources management (IWRM) in order to optimize the use of limited financial resources and capacities in the region.

18. Climate variability and change is already affecting water resources and their management in many parts of the region, as notably illustrated by the large-scale retreat of Himalayan glaciers and the growing threat of the rising sea level to small island countries. Adaptation to the effects of climate change and increasing climate variability must therefore be given the highest priority in order to secure limited water resources and to reduce the impacts of water-related disasters.

19. Adaptation and mitigation can act as complementary response measures that should be combined to reduce the risks of climate change. Water plays a central and important role in adaptation to climate change, and as such must be given central priority in national strategies for sustainable development and public security.

20. The lack of firm commitment by the governments of many countries in the region, accompanied by a lack of effective policies, planning mechanisms, and monitoring of investments and results, has resulted in inadequate progress in meeting both the MDGs and national socio-economic goals of development. The urgent need to improve governance, including the better allocation of financial resources and more effective investment in water resources management, is thus a priority issue for regional cooperation in the region.

21. In short, the Asia-Pacific's water challenges are of massive proportions. Resolving these challenges would produce substantial social, economic, and environmental benefits for the region's countries and citizens. However, achieving major progress will require strong support at the highest levels of decision-making. The 1st APWS provides an unprecedented and historic opportunity to engage in collective efforts to recognize that each of the important issues raised by the APWF in this Policy Brief are closely interlinked and thus need to be tackled in a comprehensive manner within the perspective of IWRM.

22. Unfortunately, we cannot solve all of our problems at once. The following sections of the Policy Brief will focus on three specific themes, which were identified in the Asia-Pacific Regional Document for the 4th World Water Forum (March 2006) as those most urgent for the region. The APWF firmly believes that achieving progress under these Priority Themes will also help to solve many of the other problems described above.

Priority Theme A:
Water Financing and Capacity Development

23. The Priority Theme on Water Financing and Capacity Development aims to increase targeted investments in water management in the Asia-Pacific region, in supply and sanitation infrastructure, and in human resources development in particular. The theme focuses explicitly, but not exclusively, on the efforts required for meeting the MDG targets for safe drinking water and sanitation (Target 10) and for improving water resources management in general, especially in the context of the impending impacts of climate change.

24. Improvements in the MDG indicators for income levels, hunger, communicable diseases, maternal and child mortality rates, and environmental sustainability rest on many factors. However, all have one factor in common: water. Water is essential to sustain life, provide food, and create the conditions for higher levels of health, education, and income. Investing in the water sector embodies the intent of "Human Security," which is to create systems that give people the building blocks for survival, dignity, and livelihood. Furthermore, since water is also a key element of financial development and economic growth, investing in the water sector also promotes economic growth and prosperity which can in turn lead to even greater "Human Security".

25. The Asia Water Watch 2015 study[3] estimated that an annual investment of at least $8 billion will be needed in the Asia-Pacific region over the next decade if countries are to meet the MDG targets for safe drinking water and sanitation (Target 10). The greatest proportion of these financial needs is in the South and Southwest subregions, followed by East and Northeast Asia. In terms of numbers of unserved people, the largest numbers are found in India and the People's Republic of China. Because the Asia-Pacific bears the majority of the world's poor, achieving Target 10 in this region would make a significant contribution towards attaining the global target. The achievability and affordability of meeting Target 10 in the Asia-Pacific raises an interesting possibility: political leaders can afford to confidently set more ambitious targets than the MDGs and advance their countries towards greater levels of social and economic development. It is not a matter of possibility, but a matter of willingness[4].

26. An investment in the water sector is an investment in all the MDGs, not just Target 10. The impact of water sector investments directly targeted at poor consumers is anything but subtle. Safe water supplies immediately improve people's health and save them time, which they can use for studying or improving their livelihoods, so that they can earn more, eat more nutritiously, enjoy healthier lives, and contribute to the local (and national) economy. In addition, improved sanitation protects the poor from socially and physically degrading surroundings, health risks, and exposure to dangerous environmental conditions.

© Asia-Pacific Water Forum

27. Significant investments are also needed in irrigation services, river basin management, flood management and mitigation, and wastewater management. Investments in water are crucial to meet the broader MDG targets of reducing poverty, hunger, child and maternal mortality, and the incidence of major diseases, and to improve environmental sustainability. Clearly, this is a major challenge, and one that is complicated by a wide range of governance, institutional, social, environmental, and political issues.

28. There is also a need to drastically reduce the vulnerability of populations to water-related disasters. In the last century, the Asia-Pacific region accounted for 91% of the world's total deaths due to natural disasters and 49% of the world's total damage due to natural disasters[5]. With regard to climate change, the number of typhoons and cyclones has been increasing as have the number and expanse of drought-affected areas. Disaster risk management and the expansion of facilities for water resources development, water supply, irrigation, wastewater treatment, and recycling, need to be promoted with community participation.

29. In general, it is clearly understood that the principal challenge lies not in the technological aspects (hardware component) of water supplies and sanitation, but more in the soft components, especially: (1) leadership and commitment; (2) the need for an appropriate policy environment, and legal and regulatory framework (including local government financing); (3) capacity development of service providers; and (4) innovative financing mechanisms that may prove attractive to potential investors, especially from the private sector, which would increase water sector investment while ensuring the active participation of all affected stakeholder groups – not only from the water sector but also from health, education, and other sectors concerned.

Key Messages and Recommendations under Priority Theme A

I) Water problems of developing countries are manageable

30. A key message emerging from our experience is that the region currently has a reasonable level of knowledge, experience, and technology to solve its water problems. The solutions may differ from one country to another, and even from one part of a country to another, because of differing physical, climatic, economic, social, environmental, legal, and institutional conditions. However,

the knowledge base to structure proper solutions is currently available from the lessons learned in the Asia-Pacific region. Still, to formulate and implement solutions in a timely, cost-effective, pragmatic, and high-quality manner will require significant additional investment funds, strong political will, and appropriate capacity development at all levels.

II) Give water and sanitation a higher priority on national, regional (Asia-Pacific), and international agendas

31. Although the world community is committed to working together to meet the MDGs with the current trend in progress of compliance, it is unlikely to meet Target 10, generally, in the Asia-Pacific. Attaining Target 10 will be important in improving the socio-economic and environmental conditions of developing countries in the region. However, the firm and continuing commitment of concerned decision-makers and political leaders is necessary. Hence, it is time to put water and sanitation high up on national, regional, and international agendas.

III) The principal challenges will not be technological – the hardware of water sector development – but rather the soft issues[6]

32. Decision makers need to address a range of issues in the following four core areas that reflect the key constraints identified in the UN MDG Task Force report:

33. **Policy, legal, and regulatory reform.** Reform is an essential condition for sustainable and effective change. Governments should ensure that a supportive environment exists for sustained efforts in the sector and should clearly define the roles, rights, and responsibilities of all actors. Special attention should be given to defining policies that target the specific needs and opportunities of the poor, for enhancing human security through improved access to water supply and sanitation.

34. **Planning and technology choices.** The range of technology and management choices must be broadened to include the development of innovative, low-cost technical approaches that can be implemented in poor communities. How technology can be used by communities and the participation of those communities in the planning process should always be taken into consideration. For water supply, this should include approaches such as small pipe water networks[7] for urban fringes and densely settled rural areas, as well as innovative technologies to ensure adequate water supplies. The performance of existing water supply systems should also be improved through the rehabilitation of facilities. For rural sanitation, a scaling-up of ideas for approaches that accomplish both the safe use of wastewater and excreta, and ecologically sustainable sanitation needs to be considered. With regards to urban sanitation, the promotion of appropriate community-based solutions[8] in slums should be considered, in addition to upgrading and rehabilitating the ordinary sewerage systems to cope with rapid population growth and increasing water pollution.

© Asia-Pacific Water Forum

35. **Financing mechanisms.** Investment environments (especially those that encourage small private sector investments) and effective cost recovery mechanisms should be made accessible. Two aspects of this issue are particularly important: (1) the establishment of more effective and diverse credit and financial management systems that are accessible and affordable to the poor, and (2) the development of a regulatory regime that encourages investments by the private sector, and especially by local small-scale entrepreneurs. In this respect, the introduction of Public-Private Partnerships (PPP) schemes and financial mechanisms for local water operators should also be considered.

36. **Institutional reform and capacity development.** Reform should include capacity building[9], the introduction of more appropriate management systems, and more effective institutional coordination between all players. This is particularly the case among government agencies, but there is also a need to build better links between government, the private sector, civil society, and community-level organizations. Three critical gaps need to be addressed: (1) strengthening decentralization processes and building the capacity of local government agencies; (2) supporting the development of local private sector providers by developing effective market systems; and (3) reducing fragmentation and clarifying uncertainties over mandates that often characterize the water supply and sanitation sector, as well as the management of water resources.

IV) More investment in the water resources sector is necessary to achieve MDG Targets in the Asia-Pacific. It is not a matter of possibility but a matter of willingness

37. As explained above, a regional investment of just $8 billion a year would ensure that the Asia-Pacific meets Target 10 and continues to expand coverage beyond 2015. The different development needs of urban and rural areas in relation to water need to be taken into account when formulating developmental strategies. For instance, urban areas, particularly slums, need water supply and sewage water treatment, poverty reduction, and flood control, while rural areas need water supply, sanitation, irrigation, and water resources development.

38. While internal funds may not be available in some developing countries, financial support is available through several external support agencies and institutions, as well as through the private sector. At the same time, appropriate financing mechanisms, such as proper cost-sharing between governments and users and the utilization of the additional resources accrued from recent rapid economic growth, for much needed works such as sewerage and sanitation, should also be created.

39. Money is available[10]. What is needed now is firm political will and commitment to invest in water and sanitation, to preserve and upgrade the environmental condition of our scarce water resources, and to show leadership towards proper reforms and real capacity development in the region.

Priority Theme B: Water-Related Disaster Management

40. The Priority Theme on Water-Related Disaster Management aims to create new momentum towards commitment and leadership in order to strengthen and ensure the effectiveness of our efforts to achieve a positive change in the way water-related disasters are managed in the Asia-Pacific region.

41. The correlation between poverty and population density versus the toll of human casualties clearly reveals how poverty, societal inequality, and a lack of political commitment to adapt to changing risks, also create additional elements of vulnerability to water-related disasters. Health issues, especially the spread of epidemics and water-borne diseases during disasters, remain a major challenge for many communities, particularly for those situated in slums.

42. As recently highlighted by the IPCC, the probability of increasing extreme climatic events such as floods, droughts, and coastal flooding induced by global warming is likely to further aggravate the impact of disasters in the region. Heavy precipitation events, which are very likely to increase in frequency, will augment flood risk. In the course of the next century, water supplies stored in glaciers and snow cover are projected to decline, reducing water availability in regions supplied by meltwater from major mountain ranges, where more than one-sixth of the world's population currently lives. Densely-populated and low-lying areas where adaptive capacity is relatively low, and which already face other challenges such as tropical storms and local coastal subsidence, are especially at risk. The populations affected will be largest in the mega-deltas of Asia and Africa, while small island countries are especially vulnerable[11].

43. Comprehensive and integrated disaster management schemes, such as the integration and best mix of both structural and non-structural measures for disaster management, must be sought for optimizing the use of limited available resources. Structural measures such as levees, water/sediment regulation facilities, and rainwater harvesting need to be steadily implemented, as they would increase the level of physical protection from certain disasters. At the same time, the potentially significant environmental impact of any of these measures would need to be factored into the decision-making processes, together with socio-economic factors, with the understanding that final decisions on such developments are matters of societal choice.

44. From a global perspective, the MDGs, the Hyogo Framework for Action 2005-2015 (HFA), and the Hashimoto Action Plan, together provide a set of strategic goals and global actions for substantially reducing the effects of disasters on lives and livelihoods over the decades to come. In line with the above, it is vital that national and local development plans be formulated to mainstream water-related disaster management, including climate variability and change, into national land and water management policies and practices and that this mainstreaming process be backed by proper levels of investment.

Key Messages and Recommendations under Priority Theme B

V) Integrate Water-related Disaster Risk Reduction (DRR) into national development plans, recognizing adaptation to increasing risks from climate change as a "highest" priority issue

45. Considering that the majority of natural disasters in the Asia-Pacific region are water-related, especially those caused by tropical cyclones[12], the current trend of increasing loss of life and livelihoods triggered by water-related disasters, and floods in particular, can be regarded as a major impediment to sustainable development and poverty reduction. The expected increase in the severity of water hazards due to climate change will further exacerbate the damage, both in human and economic terms, especially when combined with existing vulnerabilities such as ill-performing social, cultural, political, and management structures.

46. Adaptation to the possible increase in risks from global warming is a new and common challenge to all countries in the region and thus needs to be recognized as a "highest" national priority that must be tackled at various levels and by different sectors in a concerted manner.

47. The integrated approach to disaster management needs to involve various sectors such as urban and land use planning, societal development, industry, education, agriculture, and environmental protection. Therefore, the establishment of a comprehensive policy for managing water-related DRR is a necessity which should be integrated into national and local development plans for water-related DRR. In this respect, a message such as this: "Ensure that DRR is a national and local priority with a strong institutional basis for implementation"[13] should also be declared as a high level political agenda.

48. In tackling water-related DRR, the involvement of all stakeholders is a must. But above all, national governments should continue to assume central responsibility for national security and peoples' well-being by establishing national strategies from a longer and wider perspective, ensuring necessary resources, and putting various preventive and adaptive measures into practice. Such strategies would address the need to considerably increase investments in water-related risk management, which is currently very low in most developing countries, to a desirable level as illustrated by several good examples in the region[14].

VI) Recognize the importance of IWRM for water-related DRR and the need to strengthen comprehensive structural and non-structural measures

49. In order to ensure the optimized use of water resources it is important to look at the whole water regime, which means that both scarce and excessive water management must be done in a concerted manner. In recent years, IWRM has been widely and internationally recognized as a desirable approach towards water issues. IWRM is a broad concept promoting the integrated management of water in a sustainable and equitable manner and thus encompasses a wide variety of sectors such as physical, geographical, socio-economic, and cultural domains. *The IWRM approach, therefore, must also be applied to water-related DRR.*

50. In this context, we should note the message delivered by Ministers of the Asia-Pacific region to "promptly strengthen the comprehensive efforts from both hard and soft sides, based on the recognition that the measure is an important element for integrated water resource management, against water-related disasters."[15] Attempts to tackle water-related disaster management from an IWRM perspective are already underway[16]. The adoption of the IWRM approach would ensure various forms of integration (e.g., land and water management, flood and drought management, structural and non-structural measures, etc.). This also contributes to the most effective use of limited resources and capacities in the region.

VII) Establish national and local goals/targets for water-related DRR, taking the impacts of climate change into consideration

51. It generally takes a long time for water-related DRR efforts to show visible results in terms of national sustainable development. Therefore, central governments, local governments, and institutions which focus on disaster management should identify clear-cut goals/targets in water-related DRR for reducing the loss of life and livelihood, reflecting social and economic conditions of the communities at risk. Establishing national and local goals/targets is thus important as it provides a basis leading to concrete actions. In this sense, we appreciate the recommendation issued by the high-level water-expert panel to: "Establish, with unified political will, a clear-cut global-level target that articulates the direction for global actions for reducing the loss of life and livelihood caused by water-related disasters."[17]

52. Setting proper DRR targets is also a new challenge posed by climate change, under which decisions need to be made on how to assess the magnitude of future hazards and how to select adaptive management strategies in uncertain situations.

VIII) Develop preparedness indices for water-related DRR for the Asia-Pacific region

53. The MDGs and the HFA, among other international and regional frameworks, are undoubtedly strong vehicles and provide momentum to create unified efforts towards making sustainable progress in water-related DRR.

54. In following-up these frameworks, the development of flood and drought preparedness indices will enable central and local governments to monitor the performance of their achievements and the effectiveness of actions and policies undertaken to reduce the impact of water-related hazards. The utilization of indices will ensure facilitation of the positive spiral of national and local preparedness building up in the Asia-Pacific region. These indices must be adaptable in order to address community preparedness and vulnerability reduction at each stage of the disaster management cycle.

© Ruedee Arunkhajohnsak

55. The indices for water-related disaster preparedness are expected to help initiate a platform for the creation of standardized procedures for water-related disaster preparedness, which could be developed into an international standard.

IX) Develop water-related disaster warning systems and human capacities

56. Experience shows that effective prevention strategies would greatly contribute to reducing both human and economic losses, as opposed to investments in response and recovery. In this context, the development of well-functioning Forecasting and Warning Systems (FWS) is a prerequisite for communities' disaster preparedness and adaptation. To this end, it is necessary to undertake a consolidated review on the currently available FWS in the Asia-Pacific region and develop appropriate FWS within the parameters of available technologies and resources. The utilization of real-time satellite data could prove useful in the Asia-Pacific region where the availability of real-time data is scarce and thus needs to be promoted.

57. Community capacity development also plays a key role in the successful implementation of various measures in localities where disaster management operations are mostly carried out, in order to build community resilience to water-related disasters. Advanced high-technology can only work when proper local/community capacity development and a culture of disaster prevention, such as community-based disaster management, are implemented. To this end, various levels of education for flood management unique to the Asia-Pacific region must be provided. This will enhance local awareness in order to cultivate qualified disaster managers.

Priority Theme C: Water for Development and Ecosystems

58. The objective of the Water for Development and Ecosystems Theme is to secure water for long-term, multi-sectoral needs, generating both developmental and environmental dividends.

59. In March 2005, the governments of the Asia-Pacific region adopted a new approach to sustainable development: *Green Growth* which is designed to reconcile seeming conflicts between commitments to environmental sustainability and poverty reduction within the framework of the MDGs by moving away from an emphasis on economic growth with some attention to the environment, to a more integrated thinking in which development enhances sustainability instead of threatening it. Decision-makers must now commit to specific, tangible and achievable objectives. The application of emerging approaches and tools to water management can drive real progress.

60. Decision-makers should invest in ecosystems as development infrastructure that needs to be maintained, restored, monitored, and managed. Investment in water infrastructure for water supply, sanitation, and flood control can be more efficient if ecosystem services provided by aquatic ecosystems, upstream watersheds, and agricultural landscapes are recognized and preserved. Because of the preponderant role of agriculture in water usage, rethinking agricultural water management to improve its productivity and decrease its environmental footprint has the widest scope for improvement. This will be a key area of reform, investment, and capacity building to reach the MDGs in the region.

Key Messages and Recommendations under Priority Theme C

X) Implement good governance at the basin scale

61. The need for ecosystem approaches to development has resonated with many decision-makers across the region, yet the implementation of ecosystem approaches, particularly at the river basin

scale, has not been effective. Similarly, the decisive role that governance plays in determining the outcomes of development is increasingly recognized by all sectors of society. In much of the region, governance at the basin level is weak because political and natural boundaries do not coincide. Rights over water are not clear, and the deterioration of water quality has proven difficult to control. However, there has been much progress in establishing river basin organizations which strive to manage resources at levels appropriate to the hydrological cycle and the functions of water-based ecosystems. These must be replicated. The following approaches for improved water governance are suggested:

62. **Integrate Principle 10 of the Rio Declaration into water resources policy and enhance coordination by increasing stakeholder representation and participation in decision making processes.** The governments of the region have already committed to, and should now take decisive action towards, and monitor, enhancing governance in development through Principle 10 of the Rio Declaration, which calls for ensuring access to information, participation and justice in environmental matters in decision-making processes. National decision-makers should provide the legal and policy framework for a diverse range of institutions – including river basin organizations, networks, coalitions, and dialogue platforms – that can provide necessary inputs from the different sectors of society and improve coordination among stakeholders.

63. **Integrate groundwater into basin-level management.** Groundwater resources are an important component of many of the region's critical ecosystems and food systems. Governments in the region must prioritize groundwater resources within plans for sustainable development. Integration of groundwater into water policy will require a major departure from the *status quo*, requiring much more thorough data and analysis and revision of development policies that create the dynamics for groundwater depletion.

64. **Enhance transboundary collaboration through more open and informed dialogue and adopt and implement regional frameworks to reduce ecological and social threats:** The Asia-Pacific governments must further commit to engaging in meaningful dialogue, institutionalizing protocols and mechanisms for information sharing and consultation on joint development and action plans on transboundary issues, while making monitoring information available to society at large. International and regional cooperation bodies should support these efforts. Where large-scale water management schemes (such as hydroelectric power and irrigation) may have transboundary impacts, institutions in shared basins need to address the political complexity of international water and the transboundary tensions created.

XI) Adopt practical tools to improve economic, social and environmental outcomes of water management

65. There is an urgent need for more holistic approaches to achieving successful outcomes in the environmental, economic, and social spheres. Water management plans must recognize that ecosystems themselves require certain levels of water flow in order to remain productive and provide for the needs of all stakeholders in society. Likewise, ecosystem services and public goods delivered by agriculture water management systems also need water. Tools for dealing with these issues include:

66. **Adopting environmental flows (E-flows) in water development planning and policies.** Ecosystems and livelihoods are often affected when inadequate consideration is given to the potential impacts of water-related development and management strategies. An environmental flow is a water regime within a river, wetland, or coastal zone for maintaining ecosystems and their benefits where there are competing water uses. E-flow approaches assist in assessing the distribution of costs and benefits across sectors and stakeholders and can help ensure that water is allocated to the environment in development planning, especially involving large infrastructure, and should be seen as integral to sustainable water management. Governments of the region must establish legislative, regulatory, and institutional policies and capacities to enable effective quantification, implementation, and enforcement of environmental flow, recognizing the ecosystem services of agricultural systems.

67. **Developing mechanisms and markets for payment for environmental services.** While economic growth is rapid across much of the region, the distribution of benefits across societies is more uneven. The rural poor are often being asked to maintain supplies of critical ecosystem goods and services such as watershed functions. Payment for environmental services schemes have been tested in the region in upstream watersheds, but should be taken to the next step of refinement and considered also for agricultural water management. Incentives such as tax breaks, tenure, and other financial and non-financial instruments must also be considered.

XII) Re-invent and invest in agricultural water management to raise the productivity of water and decrease the environmental footprint of agricultural production

68. To meet the future food security needs and rural socio-economic aspirations of the region, pressure to develop new supply sources or increase water allocation to agriculture will continue. At the same time, pressure on agriculture to perform more effectively and efficiently and reduce its environmental impact will intensify. Water management must first focus on making better use of the water being accessed, and restore a strategic balance between rain-fed and irrigated production. Given the overwhelming dominance of irrigation and its role in securing food security, it is necessary to improve the benefits of irrigation water. Irrigation is often characterized by poor water productive efficiency, and poor services to farmers, limiting their social benefits and hampering their capacity to adapt to and benefit from changes in agricultural markets and water allocation. Policy-makers can avoid conflict between agriculture and competing sectors by providing direction and incentives for the reform of irrigation management and the upgrading of infrastructure, while delivering direct benefits for a rapidly growing population and to farming communities. Proposed approaches include:

69. **Taking more balanced strategic approaches to increase water productivity in both rain-fed and irrigated agriculture.** The potential to increase food production by boosting the productivity of rain-fed production has been neglected and must be fully realized by providing incentives for the adoption of improved soil and water management practices, and supplementary irrigation and water harvesting, resulting in more equitable investment patterns.

70. **Recognizing and managing the multiple roles of irrigation water.** In addition to food, irrigation systems in the region also provide water for farmhouses, habitats for fish and other aquatic resources, rural enterprise water supplies, domestic water, hydroelectric power, and navigation. It also supports important cultural values that are essential for local wellbeing and livelihoods. Ecological benefits include flood control, groundwater recharge, water purification, biodiversity conservation, and climate adjustment. Policies that recognize and promote the multi-functionality of irrigation water can improve food security, health and sanitation of local communities, and benefits to society as a whole.

71. **Modernizing irrigation systems' management to enhance the welfare of farming communities, environmental sustainability of irrigation, and allowing reallocation of water to other uses and users.** Irrigation sector reform should be reoriented to focus on the professionalization of management and service provision and the accountability of service providers to users, satisfying the needs for representation and empowerment along with performance objectives and market-based instruments. This poses many questions regarding the capacity and willingness of the institutions to deliver services on the basis of demand rather than supply. In many developing countries with large irrigated sub-sectors, there is a continued reliance upon supply management approaches with little attention to the management of existing assets. This trend needs to be reversed if agriculture is to continue to account for its water withdrawals. Reforms will require improving both physical structures and management institutions. Investments should be based upon an understanding of the changing demand in irrigation services for farmers, multi-functionality, and the need to improve the environmental performance of the systems in a river basin management context.

72. **Investing in capacity building and mandating benchmarking systems.** Bringing about this shift in irrigation management will require considerable investment in capacity building at all levels, including engineering and operations. Countries also need to adopt benchmarking systems to monitor the effects of policy changes and improve sectoral performance and quality of investment.

73. **Adopting good practices to reduce the environmental footprint of agricultural production.** In order to counter the environmental impacts that have resulted from the intensification of agricultural production and exploitation of natural resources, policies must be revised to promote conservation agriculture, integrated pest management, integrated plant nutrition, sustainable forest management, efficient processing and marketing processes, and environmentally friendly livestock production.

XIII) Protect and restore urban environments

74. Improving the health and function of urban aquatic systems is vital for ensuring health, sanitation, and overall quality of life in fast-growing urban areas. In the urban setting, the protection of valuable systems should be coupled with the restoration of degraded systems. Concerted efforts to improve waterways and bodies of water are an important strategy for contributing to human development in cities.

XIV) Prepare for climate change impacts on water availability and adopt mitigation and adaptation strategies that provide benefits now

75. Government responses to climate change must focus on reducing human vulnerability while protecting and restoring ecosystems. Effective responses will require improved governance and concerted efforts to build society's capacity to adapt, as the basic assumptions about food production, flood protection, and resilience to drought are challenged by climate change. Improved water and ecosystem governance is essential to develop equitable strategies. Specifically, governments need to prepare for changes in the flow regimes in continental basins, rising sea-levels, and changes in ecosystem dynamics. The major policies, strategies, and techniques available for re-inventing agricultural water management would actually also reduce emissions or sequester carbon and/or represent adaptation options for vulnerable livelihood systems. Creating markets for environmental services can also promote the adoption of adaptive practices, such as the management of watersheds, natural parks and protected areas, the rehabilitation and sustainable management of grasslands, and reforestation and afforestation.

Guiding Principles to Accelerate Progress

76. The following three key messages (XV, XVI and XVII) emanated from the Key Result Area on increasing local capacity (KRA2) and were developed through cross-fertilization with the Priority Themes. These describe general approaches that should guide future actions in the region.

XV) Decentralization with empowerment

77. In many countries in the Asia-Pacific, the delivery of basic water services has been delegated to local authorities. However, there is still a big gap between this decentralized responsibility and the current capacity for supporting the role of local authorities. National governments should focus on strengthening local level delivery systems. Local authorities must be equipped with an enabling environment, allowing them to ensure sustainable service provision for all. This would include the further development of existing capacities to plan and coordinate implementation and the provision of the necessary institutional, technical, and financial support. The political will to assume responsibility must be harnessed and capacities must be strengthened to ensure effective, sustainable, and efficient service delivery systems and good governance at the local level.

XVI) Partnership approach between local authorities, civil society (NGOs and community-based groups) and different service providers (public and private)

78. Local authorities have generally been expected to ensure basic services. The rights-based approach demands that the state be expected to promote, respect, and fulfill the right of local populations to receive the services. However it is not expected that local authorities provide the services by themselves. The key is to be open to multi-stakeholder participation and community-led initiatives. Collaboration is crucial at different levels in order to achieve the mutually accepted goals of sustainable service delivery for all, targeting especially the poor and other marginalized groups of men, women, and children, the sick, and the elderly. Participation in planning, budgeting, technology choices, performance goal setting, and monitoring are some of the key areas that need to be opened up for collaborative action.

XVII) Capacity development as an important area of investment

79. The delivery of basic water services is not just about infrastructure. However, most Asia-Pacific governments fail to provide resources for capacity development which ensure that the knowledge, skills, and behavior changes are in place. While there is a wealth of knowledge, experience, and expertise in the region, there is still limited access to the right information, materials, and tools; insufficient funds for supporting the knowledge transfer process; and insufficient or inadequate human and organizational resources. Adequate investment in capacity development enhances the ability to work with informed partners, work at scale, replicate good practices, and innovate. It also enables local actors to provide the necessary support system to follow through and ensure the continuity of an environmentally sustainable service delivery system.

Key Recommendations for the Way Forward

80. These final recommendations of the Policy Brief emerged from a combination of several complementary recommendations that in turn emerged from the Key Result Area on developing knowledge and lessons (KRA1) and the Key Result Area on monitoring investments and results (KRA4) as well as from the Priority Themes. They encompass a number of specific concrete initiatives which are listed in Annex A.

XVIII) Create a platform of scientific information to support decision-making

81. There is a fundamental mismatch between the scales of information management and decision-making. Most data indicators of water availability and scarcity are collected at the national level, and feed into national development planning. But water availability and quality are tangible at the ecosystem level, be it in sub-national or transboundary basins. Policies to address these issues at the relevant level need to be informed by data on a corresponding scale via indicators that integrate the water required by an ecosystem to maintain the flow of goods and services.

XIX) Create a platform for knowledge coordination, knowledge sharing, and knowledge management at the regional and national levels to increase access to important information, skills, and other support services, in order to improve capacities and efficiencies among implementers of water management programs and services

82. While there are a lot of valuable experiences and information around the region, access of local implementers to this kind of knowledge base is not easy. Capacity development in the region should address the needs of the different stakeholders who are key in implementing policies at various levels. There must be a systematic way of knowledge sharing and skill development based on demand.

XX) Create a platform for a set of sector-level collective monitoring systems at the regional and national levels

83. Monitoring the entire water sector at the national level is essential in order to support the formulation of effective policies for socio-economic development and environmental management. For this purpose, not only are mechanisms for monitoring necessary, but a proper system linking the results of monitoring to policy making is also essential.

84. In view of the complexity of the water sector at the national level, it is likely that a set of sector-level monitoring programmes would prove to be most effective:

- **Water supply and sanitation**[18]. The purpose of the system should be to collect and analyze data on sector financing and investments and the outcomes of investments in order to provide information for guiding the development of national sector policies as well as financial and technical support programmes.

- **Agricultural water management.** Monitoring information available in the sector has pointed to the poor performance of many irrigation systems and a major reduction in investment for irrigation infrastructure in the region. National governments need to monitor investment and its results in irrigation at both the national and regional levels in order to adjust investment strategies to meet food securities in basic food staples, adjusting for national self-sufficiency targets and availability and trends in international markets.

- **Ecosystem management.** Governments are urged to establish, with the assistance of international organizations, a monitoring system for the status of rivers and ecosystems in their respective countries and to mobilize the participation of local authorities in developing plans for rehabilitation and restoration.

- **Water-related disaster management.** Governments are urged to adopt standardized methodologies of assessing the socio-economic impacts of water-related disasters to form a common basis for collaboration among countries and international organizations, including financing agencies, as part of a strategy towards monitoring.

- **Integrated water resources management.** Monitoring of investment and results of the entire water sector at the national level is essential. The most effective of the possible options for such monitoring mechanisms are identified as follows:

  - **Option 1:** Establishment of a permanent mechanism for overall monitoring of investment and results at the national level for regular monitoring at appropriate frequency

  - **Option 2:** Establishment of a policy on the overall monitoring of investment and results, through which *ad hoc* mechanisms could be established at specific times, with the appropriate frequency

85. In order to ensure synergy of regional efforts on strengthening the monitoring of investment and results, it is recommended that these regional efforts be integrated into the process of institutional strengthening of regional cooperation through APWF. For this purpose, the following courses of action are recommended:

- Establishment of a network of international organizations and financing institutions, in coordination with existing programmes, in order to support the process of strengthening the monitoring of investment and results in the region, and to study implications of findings from the monitoring of investment and results for more effective regional cooperation in water resources management, especially for consideration at the regular Summit.

- Establishment of a Ministers' Council for Water of the Asia Pacific in order to assist developing countries to strengthen their capacity in policy development and to enhance the benefits of regional cooperation.

- Integration of the findings of the monitoring of investment and results in water resources, with the operations of the proposed Central Water Knowledge Hub for Capacity Development

Endnotes

1. WHO/UNICEF Joint Monitoring Programme. www.wssinfo.org
2. IPCC 4th Assessment Report. 2007.
3. This study was commissioned by ADB, World Health Organization (WHO), United Nations Development Programme (UNDP), and United Nations Economic and Social Commission for Asia and the Pacific (UNESCAP).
4. ADB. 2006. Asia Water Watch 2015. http://www.adb.org?Documents/Books/Asia-Water-Watch/default.asp
5. ADB. Technical Assistance Report: Supporting the Asia-Pacific Water Forum
6. For irrigation, infrastructure does matter. Not only is it in poor condition, but its design parameters are generally obsolete and need to be modernized to support new software arrangements and improved service delivery.
7. http://www.adb.org/media/printer.asp?articleID=10235
8. These include community toilets, and condominium sewerage connection systems.
9. Including the establishment of a central water knowledge Hub for Capacity Development in Singapore.
10. To address these concerns, ADB in 2006 announced a significant Water Financing Program (WFP), intended to double its investments in water to over $2 billion annually, and to increase its technical assistance for preparing feasibility studies and long-term capacity building programs across the sector. ADB has also established a Water Financing Partnership Facility to support WFP. Currently targeted at $100 million, the facility will provide resources that augment knowledge development and capacity building.

    Japan is also expected to play an important role. Japan has been the most active donor in water sector in this region, allocating 20% of ODA loan aid and 7% of grant aid to the solving of water issues, and is committed to continue playing an active role. In addition, the new JICA (which will be established in October 2008 through the merger of JICA and JBIC), and the consolidation of technical cooperation, grant and loan aid of Japan's ODA, will bestow aid modalities for both investment and capacity development. It is, therefore, expected that the new JICA will be able to address issues related to water with more flexibility and mobility.
11. IPCC 4th Assessment Report. 2007. http://www.ipcc.ch/SPM13apr07.pdf
12. According to a 2006 study by UNESCAP and the basis of EM-DAT, tropical cyclones accounted for 70% of the number of people killed and about 90% of the total economic damage caused by water-related disasters.
13. Hyogo Framework for Action, World Conference on Disaster Management, Hyogo, Japan. 2005. http://www.unisdr.org/eng/hfa/hfa.htm
14. Experiences in Japan showed the desirable investment level to be 1% of GDP. Experiences in Malaysia and the Republic of Korea illustrated a consistent increase from 0.01% to 0.04% in past decades.
15. 6th Ministers' Forum on Infrastructure Development in the Asia-Pacific, Beijing. 2007. http://www.mlit.go.jp/kisha/kisha07/01/010830_2/02.pdf
16. International Flood Initiative. http://unesdoc.unesco.org/images/0014/001463/146389 and the WMO/GWP Associated Programme on Flood Management. http://www.apfm.info
17. Hashimoto Action Plan: Compendium of Actions, United Nations Secretary General's Advisory Board on Water and Sanitation. 2006. www.unsgab.org/docs/HAP_en.pdf
18. In order to avoid duplication of effort and promote harmonization in the collection and analysis of data and information, the system should recognize and receive regular inputs from ongoing global and regional sector monitoring and reporting programmes such as the WHO/UNICEF Joint Monitoring Programme on Water Supply and Sanitation, UN Water's annual sector monitoring report, and others.

Annex: List of Expected Concrete Initiatives to be Launched at the 1st Asia-Pacific Water Summit

Concrete Initiatives	Implementing Organization
PRIORITY THEME A	
Financing Water and Sanitation and Strengthening Capacity Development	Asian Development Bank (ADB), Japan Bank for International Cooperation (JBIC), Japan International Cooperation Agency (JICA)
PRIORITY THEME B	
Regional Knowledge Hub for Water-related Disaster Risk Reduction (ICHARM)	The International Center for Water Hazard and Risk Management (ICHARM)
Adaptation to Climate Change concerning the impacts of glacier-melting in the Hindu Kush-Himalayan (HKH) region	International Centre for Integrated Mountain Development (ICIMOD)
Pacific Partnership Initiative on Sustainable Water Management	Pacific Islands Applied Geoscience Commission (SOPAC)
PRIORITY THEME C	
Asia Pacific community practice on environmental flows for water resources management at policy, river basin and local levels	The World Conservation Union (IUCN), Network of Asian River Basin Organization (NARBO), International Water Management Institute (IWMI)
Re-inventing irrigation and agricultural water governance in the Asia Pacific to reach the Millennium Development Goals: a regional programme to support action at all levels through knowledge sharing, capacity building and implementation	Food and Agriculture Organization of the United Nations (FAO), IWMI
KEY RESULT AREA 1	
Initial Knowledge Hubs	Public Utilities Board Singapore (PUB Singapore), United Nations Educational, Scientific and Cultural Organization (UNESCO)
Asia-Pacific Water Museum	National Science Museum of Thailand, UNESCO, Japan Water Forum (JWF)
KEY RESULT AREA 2	
Capacity Building Hubs and Policy Support for Increased Local Capacities	United Nations Human Settlements Programme (UN-HABITAT) in collaboration with Streams of Knowledge (SOK)
KEY RESULT AREA 3	
Asian Water & Development Outlook	ADB
Asia-Pacific Water Documentary	Asia-Pacific Broadcasting Union (ABU), ADB, JWF
KEY RESULT AREA 4	
Regional network of water ministers to strengthen regional cooperation in water resources	United Nations Economic and Social Commission for Asia and the Pacific (UNESCAP)
Regional water task force to support the strengthening of water resources management with emphasis on monitoring of investment and results in the water sector	UNESCAP
OTHER INITIATIVES	
Launching of the International Year of Sanitation 2008 in Asia and the Pacific	United Nations Department of Economic and Social Affairs (UNDESA), UNESCAP, JBIC and ADB
Water Web Project on Google map & earth	JWF, Water Web Alliance
A Call to Action: Asia-Pacific Businesses to Address the Water Challenge	United Nations Global Compact
Aral Sea Basin Public Water Council	Executive committee of the International Fund for Saving the Aral Sea (EC IFAS), Interstate Commission for Water Coordination of Central Asia (ICWC), Interstate Commission for Sustainable Development (ICSD), Global Water Partnership of Caucasus and Central Asia (GWP CACENA)

Asia-Pacific
Water Forum

c/o Japan Water Forum
5th Fl., 1-8-1 Kojimachi Chiyoda-ku, Tokyo 102-0083 Japan
Tel: +81 (0)3-5212-1645 Fax: +81 (0)3-5212-1649
http://www.apwf.org

Message From Beppu

別府からのメッセージ

Asia-Pacific Water Forum

1st Asia-Pacific Water Summit Beppu Japan 2007

We, the leaders of the Asia-Pacific, coming from all sectors of our societies and countries, meeting at the historic inaugural Asia-Pacific Water Summit, in the beautiful city of Beppu, in the hospitable Oita Prefecture of Japan, do hereby agree to:

- *Recognise the people's right to safe drinking water and basic sanitation as a basic human right and a fundamental aspect of human security;*
- *Reduce by half the number of people who do not have access to safe drinking water by 2015 and aim to reduce that number to zero by 2025;*
- *Reduce by half the number of people who do not have access to basic sanitation in our region by 2015 and aim to reduce that number to zero by 2025, through the adoption of new and innovative sanitation systems that are not as water reliant as current methods;*
- *Accord the highest priority to water and sanitation in our economic and development plans and agendas and to increase substantially our allocation of resources to the water and sanitation sectors;*
- *Improve governance, efficiency, transparency and equity in all aspects related to the management of water, particularly as it impacts on poor communities. We recognise that while women are particularly vulnerable, they are also resilient and entrepreneurial, hence, should be empowered in all water-related activities;*
- *Take urgent and effective action to prevent and reduce the risks of flood, drought and other water-related disasters and to bring timely relief and assistance to their victims;*
- *Support the region's vulnerable small island states in their efforts to protect lives and livelihoods from the impacts of climate change;*
- *Exhort the Bali Conference to take into account the relationship between water and climate change, such as the melting of snowcaps and glaciers in the Himalayas and the Pamirs, and rising sea levels, which are already having an impact on some countries in the region;*

- *Establish concrete goals for the 2008 Toyako G8 Summit to:*
  - commit to support the developing countries to achieve their MDG targets on water and sanitation; and
  - take immediate action to support adaptation to climate change by developing countries;

- *Empower a high-level coordinating mechanism in our cabinets and where possible, appoint a minister in charge of water to ensure that all issues related to water and sanitation would be dealt with in a holistic manner;*

- *Respect and strengthen the region's rich history of water-centered community development, including the rehabilitation of urban waterways and protecting the environmental integrity of rural watersheds; and*

- *Work together with other like-minded institutions, entities and individuals in order to achieve our collective vision of water security in the Asia-Pacific region.*

We will support the Policy Brief as prepared by the Asia-Pacific Water Forum family.

We encourage all governments to make all efforts to implement its recommendations.

We have the will and courage to realise our vision.

NOTE:
This Message from Beppu was unanimously endorsed by the participants of the 1st Asia-Pacific Water Summit, which was held in Beppu, Japan, on 3-4th December 2007, attended by ten Heads of State and Government, 31 Ministers, and representatives from over 36 Asia-Pacific countries and regions.

Message From Beppu

ANNEX 1

Goverment Delegates*

Republic of Armenia
Hon. Mr. Aram Khachik Harutyunyan
Minister of Nature Protection

Australia
H.E. Mr. Alistair Murray McLean
Ambassador
Australian Embassy in Japan

Republic of Azerbaijan
Hon. Dr. Husseingulu Seyid Baghirov
Minister of Ecology and Natural Resources

Kingdom of Bhutan
H.E. Lyonpo Dr. Kinzang Dorji
Prime Minister
Dasho Mr. Sherub Tenzin
Cabinet Secretary

Brunei Darussalam
Hon. Mr. Haji Brahim Haji Ismail
Permanent Secretary
Ministry of Development

Kingdom of Cambodia
Hon. Mr. Lim Kean Hor
Minister of Water Resources and Meteorology

People's Republic of China
H.E. Mr. Hu Siyi
Vice-Minister
Ministry of Water Resources

*This list includes heads of governments and ministerial level participants only.

Cook Islands
H.E. Mr. Tangata Mouauri Vavia
Minister of the Crown

Republic of the Fiji Islands
Hon. Mr. Mahendra Pal Chaudhry
Minister of Finance, National Planning & Sugar Industry
H.E. Mr. Ratu Inoke Kubuabola
Ambassador
Embassy of the Republic of the Fiji Islands in Japan

India
H.E. Prof. Sauf-ud-Din Soz
Minister of Water Resources

Republic of Indonesia
Dr. Mochammad Amron
Senior Advisor to Minister
Ministry of Public Works

Islamic Republic of Iran
Hon. Mr. Seyed Parviz Fattah Gharehbagh
Minister of Energy
Hon. Dr. Rasoul Zargarpour
Deputy Minister of Energy for Water and Wastewater
H.E. Mr. Mohsen Talaei
Ambassador
Embassy of the Islamic Republic of Iran in Japan

Japan
H.E. Mr. Yasuo Fukuda
Prime Minister
H.E. Mr. Tetsuzo Fuyushiba
Minister of Land, Infrastructure, Transport and Tourism
H.E Mr. Osamu Uno
Vice-Minister for Foreign Affairs
Hon. Mr. Junzo Yamamoto
MP: Vice-Minister of Land, Infrastructure, Transport and Tourism
Hon. Mr. Masayoshi Namiki
Vice-Minister of the Environment
Hon. Mr. Hiroaki Taniguchi
Vice-Minister for Engineering Affairs, Ministry of Land, Infrastructure, Transport and Tourism

Republic of Kazakhstan
Hon. Mr. Anatoliy Ryabtsev
Chairman of Water Resources Committee
Ministry of Agriculture
H.E. Mr. Akylbek Kamaldinov
Ambassador
Embassy of the Republic of Kazakhstan in Japan

Republic of Kiribati
H.E. Mr. Anote Tong
President
Hon. Mr. Elliot Ali
Secretary for Foreign Affairs & Immigration

Republic of Korea
H.E. Mr. Choon-Hee Lee
Vice Minister
Ministry of Construction and Transportation
H.E. Mr. Myung-hwan Yu
Ambassador
Embassy of the Republic of Korea in Japan

Kyrgyz Republic
H.E. Mr. Dosbol Nur ulu
Vice Prime Minister
H.E. Mr. Askar Asanbekovich Kutanov
Ambassador
Embassy of the Kyrgyz Republic in Japan
H.E. Mr. Sapar Isakov
Head of Department of External Communications

Lao People's Democratic Republic
Hon. Ms. Khempheng Pholsena
Minister to the Prime Minister's Office
Head of Water Resources and Environment Authority
H.E. Mr. Sithong Chitnhothinh
Ambassadoor
Embassy of the Lao People's Democratic Republic in Japan

Republic of the Marshall Islands
H.E. Mr. Phillip Kemilan Kabua
Ambassador
Embassy of the Republic of the Marshall Islands in Japan

Federated States of Micronesia
H.E. Mr. Emanuel Mori
President
Hon. Mr. Lorin Robert
Secretary (Minister)
Department of Foreign Affairs
H.E. Mr. Kasio E. Mida
Ambassador
Embassy of the Federated States of Micronesia in Japan

Mongolia
H.E. Mr. Jigjid Rentsendoo
Ambassador
Embassy of Mongolia in Japan

Union of Myanmar
H.E. Mr. Ohn Myint
Deputy Minister
Ministry of Agriculture and Irrigation
H.E. Mr. Hla Myint
Ambassador
Embassy of the Union of Myanmar in Japan

Republic of Nauru
H.E. Mr. Ludwig Scotty
President

Nepal
Hon. Dr. Ram Sharan Mahat
Minister of Finance
Hon. Mr. Gyanendra Bahadur Karki
Minister of State for Water Resources

New Zealand
Hon. Ms. Nanaia Cybele Mahuta
Minister of Customs, Associate Minister for Environment
H.E. Mr. Ian Forbes Kennedy
Ambassador
New Zealand Embassy in Japan

Niue
Hon. Mr. Mititaiagimene Vivian
Premier

Republic of Palau
H.E. Mr. Tommy Remengesau, Jr.
President
Hon. Mr. Fritz Koshiba
Minister of Resources & Development

Independent State of Papua New Guinea
Hon. Mr. Benny Allan
Minister for Environment and Conservation

Republic of the Philippines
Hon. Mr. Jose L. Atienza, Jr.
Cabinet Secretary
Department of Environment and Nature Resources
Hon. Mr. Demetrio L. Ignacio, Jr.
Undersecretary
Department of Enviroment and Natural Resources

Independent State of Samoa
Hon. Mr. Tuisugaletaua Sofara Aveau
Minister of Works, Transport and Infrastructure

Republic of Singapore
H.E. Professor Tommy Koh
Ambassador-at-Large
Ministry of Foreign Affairs
H.E. Mr. Tan Chin Tiong
Ambassador
Embassy of the Republic of Singapore in Japan
Mr. Khoo Teng Chye
Chief Executive
PUB Singapore

Democratic Socialist Republic of Sri Lanka
Mr. Mahinda Dharmasiri Bandusena
Secretary to the Prime Minister

Republic of Tajikistan
H.E. Mr. Emomali Rahmon
President
Hon. Mr. Zarifi Hamrokhon
Minister of Foreign Affairs
Hon. Mr. Erkin Rakhmatullaev
State Adviser to the President on Foreign Policy
Hon. Mr. Matlubkhon Davlatov
State Adviser to the President on Economic Policy
Hon. Mr. Gulomdzhon Bobozoda
Minister of Economic Development and Trade
Hon. Mr. Safarali Nadzhmuddinov
Minister of Finance
Hon. Mr. Yoqubzod Said
Minister of Melioration and Water Resources
Hon. Mr. Sharif Rahimozoda
Chairman of the State Committee on Investments and State Property Management

Kingdom of Thailand
H.E. Mr. Suvidhya Simaskul
Ambassador
Royal Thai Embassy in Japan

Tuvalu
Hon. Mr. Apisai Ielemia
Prime Minister
Hon. Mr. Kausea Natano
Minister for Public Utilities and Industries
Hon. Mr. Panapasi Nelesone
Secretary to Government

Republic of Uzbekistan
H.E. Mr. Bory Alikhanov
Chairman
State Nature Protection Committee
H.E. Mr. Shavkat Khamraev
Deputy Minister
Ministry of Agriculture
H.E. Dr. Mirsobit Ochilov
Ambassador
Embassy of the Republic of Uzbekistan in Japan

ANNEX 2

Sponsors, Supporting Organizations and Foundations

Sponsors

- Canon Inc.
- Coca-Cola (Japan) Company, Limited
- Grundfos Pumps K.K.
- KYUSHU ELECTRIC POWER CO., INC.
- NIKAIDOSHUZO
- SANWA SHURUI CO., LTD.
- THE OITA BANK, LTD.
- CTI Engineering Co., Ltd.
- HAKUHODO DESIGN Inc.
- Nikko Asset Management Co., Ltd.
- NIPPON KOEI CO., LTD.
- ORGANO CORPORATION
- Procter & Gamble Japan K.K.
- Tokyo Kensetsu Consultants Co., Ltd.

Canon

Coca-Cola

GRUNDFOS

KYUSHU ELECTRIC POWER CO.,INC.

むぎ100%
大分 むぎ焼酎
二階堂

iichiko

OITA BANK

CTi CTI Engineering Co., Ltd.

nikko am

NIPPON KOEI

Ecologically Clean
ORGANO

P&G

Tokyo Kensetsu Consultants

Supporting Organizations and Foundations

- Asian Development Bank
- Chugoku Construction Association
- Food and Agriculture Organization of the United Nations
- Foundation for Riverfront Improvement and Restoration
- Foundation of Hokkaido River Disaster Prevention Research Center
- Foundation Of River & Basin Integrated Communications, JAPAN
- Foundation of River and Watershed Environment Management
- Hokuriku Construction Association
- International Centre for Integrated Mountain Development
- International Centre for Water Hazard and Risk Management
- International Fund for saving the Aral Sea
- Japan Bank for International Cooperation
- Japan Civil Engineering Consultants Association
- Japan Civil Engineering Contractors' Association, Inc. Kyusyu Branch Office
- Japan Construction Information Center
- Japan Dam Engineering Center
- Japan Institute of Construction Engineering
- Japan International Cooperation Agency
- Japan Water Forum
- Kinki Construction Association
- Kyushu Construction Public Utility Association
- Pacific Islands Applied Geoscience Commission
- Public Association for Construction Service Chubu Region
- Public Association for Construction Service Kanto Region
- Public Utilities Board Singapore
- Shikoku Construction Association
- Streams of Knowledge
- The Japan Iron and Steel Federation
- The Oita Prefectural Committee for the 1st Asia-Pacific Water Summit
- The World Conservation Union
- Touhoku Construction Association
- United Nations Children's Fund
- United Nations Department of Economic and Social Affairs
- United Nations Economic and Social Commission for Asia and the Pacific
- United Nations Educational, Scientific and Cultural Organization
- United Nations Global Compact
- United Nations Human Settlements Programme

- Water Resources Environment Technology Center
- World Health Organization

In the Cabinet meeting on December 15, 2006, the Japanese Cabinet expressed agreement with the Japanese government's support of the 1st Asia-Pacific Water Summit.

ANNEX 3

Organizational Structure

Steering Committee of the 1st Asia-Pacific Water Summit

Chair: Yoshiro Mori, *President of the Asia-Pacific Water Forum/President of the Japan Water Forum/Former Prime Minister of Japan*

International board members

- Keizrul Abdullah, *Director General, Department of Irrigation and Drainage, Ministry of Natural Resources and Environment, Malaysia*
- Ahmad Mohamed Ali, *President of Islamic Development Bank*
- Margaret Catley-Carlson, *Chair of Global Water Partnership*
- Nino Chkhobadze, *Former Minister of Environment and Natural Resources Protection, Georgia/Chair of GWP-CACENA*
- Akhmetjan Esimov, *Minister of Agriculture, Republic of Kazakhstan*
- Loïc Fauchon, *President of World Water Council*
- Angel Gurría, *Secretary-General of Organization for Economic Co-operation and Development*
- Han Seung-soo, *President of Korea Water Forum/Special Envoy of the UN Secretary-General on Climate Change/President, 56th Session of the UN General Assembly*
- Noeleen Heyzer, *Executive Secretary of United Nations Economic and Social Commission for Asia and the Pacific*
- James Horne, *Deputy Secretary, Department of the Environment and Water Resources, Australia*
- Hu Siyi, *Vice Minister of Water Resources, Republic of China*
- Tommy Koh, *Chair of the Governing Council of the Asia-Pacific Water Forum/Ambassador-At-Large, Republic of Singapore*
- Haruhiko Kuroda, *President of Asian Development Bank*
- Mercedes Lentz, *Minister's Adviser of Ministry for the Environment, New Zealand*
- Cristelle Pratt, *Director, Pacific Islands Applied Geoscience Commission*
- Sulton Rahimov, *Chairman of the Executive Committee of International Fund for saving the Aral Sea*
- Quamrul Islam Siddique, *President of Bangladesh Water Partnership*
- Saif-ud-Din Soz, *Minister of Water Resources, India*

Japanese board members

- Katsusada Hirose, *Governor of Oita Prefecture*
- Yoshinori Imai, *Executive Editor, NHK Japan Broadcasting Corporation*
- Yoshihito Karasawa, *President of Japan Medical Association*
- Fujio Mitarai, *Chairman of Nippon Keidanren*
- Norio Murakami, *President, Google Japan Inc.*
- Ken Noguchi, *Alpinist*
- Sadako Ogata, *President, Japan International Cooperation Agency*
- Norihito Tambo, *Professor Emeritus of Hokkaido University and the Open University of Japan/Director of Hokkaido Historical Museum*
- Koji Tanami, *Governor of Japan Bank for International Cooperation*
- Masako Uemoto, *Former Vice President of Japanese Trade Union Confederation*
- Masae Wada, *Councilor of Housewives Association*

Asia-Pacific Water Forum

President: Yoshiro Mori, *President of the Japan Water Forum/Former Prime Minister of Japan*

Governing Council:

Chair: Tommy Koh, *Ambassador-at-Large, Republic of Singapore*

Vice Chair:
Erna Witoelar, *UN Special Ambassador for MDGs in Asia and the Pacific*
Ravi Narayanan, *Former Chief Executive of WaterAid*

Lead Organization
Priority Theme:

A. Water Financing: Asian Development Bank
B. Water-related Disaster Management: International Centre for Water Hazard and Risk Management
C. Water for Development and Ecosystems: The World Conservation Union, Food and Agriculture Organization of the United Nations

Key Result Area:

1. Developing Knowledge and Lessons: PUB Singapore, United Nations Educational, Scientific and Cultural Organization
2. Increasing Local Capacity: Streams of Knowledge, United Nations Human Settlements Programme
3. Increasing Public Outreach: Asian Development Bank
4. Monitoring Investments and Results: United Nations Economic and Social Commission for Asia and the Pacific
5. Supporting Forum and Summit: Japan Water Forum

Subregional Coordinator

- Central Asia: Global Water Partnership-CACENA/Executive Committee of International Fund for saving the Aral Sea
- Northeast Asia: Korea Water Forum
- South Asia: Global Water Partnership South Asia
- Southeast Asia: Global Water Partnership Southeast Asia
- Pacific: Pacific Islands Applied Geoscience Commission

Secretariat of the 1st APWS (in alphabetical order)

Makoto Anbe
Toshiki Aoyama
Shigenori Asai
Richard Connor
Motoo Fujiyoshi
Yoichi Fukui
Shinobu Fukumoto
Tamio Fukushima
Shuichi Hirayama
Eri Ikeda
Minoru Ikeda
Ai Isayama
Kyoko Ishiwatari
Yumio Ishii
Hitoshi Karasawa
Shohei Kawakita
Masahide Kawamura
Kunio Kawamoto
Kaori Kondo
Fumiko Kuroyanagi
Toru Miya
Aya Motojima
Ayumi Nakamura
Hideaki Oda
Naomasa Oda
Marc Overmars
Sophia Sandstrom
Marisha Shibuya
Hikaru Shoji
Yuji Takahashi
Kotaro Takemura
Michiko Takenaka
Tokuji Tomaru
Yoshie Tonohara
Masato Toyama
Kenichi Tsukahara
Yukari Tsuzuki
Noriko Yamaguchi
Taeko Yokota
Takumi Watanabe

Photographers

Shinya Hirose
Mizuho Hoshi
Satomi Matsui

ANNEX 4

Timeline of the 1st Asia-Pacific Water Summit

October 2005	Establishment of the Asia-Pacific Water Forum was proposed in Tokyo. Three Priority Themes were determined.
March 2006	Establishment of the APWF was announced in Mexico. It was proposed that Japan would host the 1st APWS.
September 2006	The APWF was launched in Manila. The venue of the 1st APWS was officially announced. The official website of the APWF and the 1st APWS was launched.
October 2006	Informal Meeting of the Leadership of the APWF Governing Council was held in Singapore. First Announcement of the 1st APWS was sent to the government in the region through embassies in Tokyo.
November 2006	Mr. Yoshiro Mori became the President of the APWF. Briefing Session on the APWF and the 1st APWS was held for embassies in Tokyo.
December 2006	Cabinet approval was issued by the Japanese Government. The Secretariat called for sponsorship of the 1st APWS.
January 2007	The Steering Committee of the 1st APWS was established with high level members.
February 2007	The 1st Governing Council Meeting of the APWF was held in Singapore. • Discussions on concept notes and progress made under each Priority Theme and KRA • Discussions on how the input from the activities of the APWF could be organised and presented at the 1st APWS • The Rules and Procedures of the APWF were adopted His Imperial Highness the Crown Prince of Japan expressed his participation in the 1st APWS.

March 2007 The 1st Steering Committee Meeting of the 1st APWS was held in Tokyo.

- Exchange of views on the programme and the selection process for the invitees of the 1st APWS

Official invitations for the 1st APWS were sent to the heads of government under the name of the Chair of the Steering Committee.

May 2007 The Secretariat determined a logo for the 1st APWS.

July 2007 The 2nd Governing Council Meeting of the APWF was held in Singapore.

- Discussions on the outcome documents
- Discussions on the 1st APWS program
- Discussions on the concrete initiatives to be presented at the 1st APWS

The Secretariat required the lead organizations to submit the draft recommendations for the Policy Brief 2007.

The Secretariat required the potential session conveners to submit the session proposals & concrete initiatives.

August 2007 The 2nd Steering Committee Meeting of the 1st APWS was held in Tokyo.

- Exchange of views on ongoing preparatory process and the preliminary draft of the Policy Brief 2007

The Secretariat organized the side event of the 1st APWS at the World Water Week in Stockholm.

September 2007 Programme Workshop on the 1st APWS was held in Bangkok for finalization of the 1st APWS programme.

October 2007 Working Group Meeting was held in Malaysia for finalization of the Policy Brief 2007.

December 2007 The 3rd Steering Committee Meeting of the 1st APWS was held for confirmation of the final programme.

The 1st APWS was held in Beppu.

ANNEX 5

Message from Dialogue between the CSOs/NGOs and Participants of the 1st Asia-Pacific Water Summit

by Peoples' Forum on Asia-Pacific Water Summit

Recommendation for Priority Theme A: Water Financing and Capacity Development

Water as a natural resource may be free, a gift of nature, but the cost of treating and delivering must be met. Governments must ensure that all have access to water and sanitation services, regardless of ability to pay. This implies a system of subsidies, from rich to poor, from urban to rural and from North to South.

International development and finance institutions, donor agencies and governments continue to promote private sector participation and full-cost recovery. However, the experience of the past 20 years indicates that market mechanisms and for-profit management fail to deliver water and sanitation service to the poor, and leave governments in the South burdened excessive and odious debt.

Public Private Partnerships (PPPs) or Private Sector Participation (PSP) have not brought significant new investment capital, but have caused serious water rate hikes and in many cases, people unable to pay are disconnected. PSP makes cross-subsidies more difficult as profits are extracted. PSP also impedes conservation efforts, as profits are based on volume consumed, not on water conserved. As well, PSP has proven incapable of meeting the needs of rural communities and of protecting the ecosystem.

Meanwhile, there are a number of good examples around the world, in which residents constructed simple reservoirs or irrigation systems by harnessing local traditional knowledge or know-how, and in the process overcame serious water shortage or groundwater depletion at minimal cost. There are also successful cases in which local communities were able to directly participate in budget allocation and implementation. As a result, local water services became more democratic, just, and efficient.

Appropriate financial policies depend on strong governance, based on transparency, accountability and participation. Citizens must be allowed and encouraged to participate in the decision-making in the sector, such that finance proposals reflect their needs and will. Many different solutions may exist if true participation is implemented.

We recommend the following:

1) Establish the human right to water and oblige international community and national governments to materialize it.
2) Cancel or radically reduce odious debt currently shouldered by developing countries.
3) Development finance should be available in local currencies in order not to be affected by currency fluctuations, and participation in PPP or PSP must not be a precondition of loans or grants.
4) Finance not- for profit small-scale drinking water and sanitation projects by grant aid.
5) Promote local community participation in budget allocation, implementation, and monitoring.
6) Harness local traditional methods, knowledge and know-how about water control and/or utilization that are more suitable to the local situations.
7) Utilize local resources including finance and personnel in water services, so that local self-sufficiency and development, as well as cost-effectiveness are attained.
8) Promote partnerships between public utilities on a not-for-profit basis, involving communities, workers and other stakeholders.

Recommendation for Priority Theme B: Water-related Disaster Management

A Knowledge Hub that reflects the traditional wisdom of local residents

It is local residents who have spent their lives with the rivers, and not only the external experts and researchers who have much information regarding disasters. There are many tips for Disaster Risk Reduction that are hidden in traditional wisdom including lore and legend. It is important to reconsider traditional techniques in a rethink on the construction of large scale facilities with heavy costs.

When establishing a Knowledge Hub, it is necessary to emphasize the traditional wisdom of local residents as well. The watershed management that passes the traditional wisdom of local residents is done. The subject and resources of the watershed management should belong to local residents.

Development related Disaster Risk Reduction (DRR)

Various problems such as environmental destruction, insufficient compensation for local residents, forced migration, and lack of transparency in projects, arise when developments such as building dams and improving river banks maintenance are implemented in the interests of Disaster Risk Reduction (DRR). We are facing the contradiction that many peoples lives and their local cultures are being threatened by development, the original purpose of which was supposed to be reduction in the loss of human lives and livlihood. Currently development information is provided in complicated documentation in English on intermediaries such as web sites. This prevents local residents from easy access to and understanding of the information.

Incase development is indispensable for DRR, we recommend the following:

1) Aiming to always put the local residents first in any development by holding timely meetings to disclose information to them, right from the preparatory stages. Disclosing information in local languages that local residents can understand. Using the intermediary of paper prints for their easy access.
2) Establishing a committee including local residents, experts, and researchers to discuss and assess the suitability and validity of the development as well as to ensure transparency.
3) Compensating affected local residents with sufficient consideration for their future livelihood.
4) Businesses and financers should have their projects assessed in line with environmental guidelines and perform their businesses with consideration for the environment.

Disaster Risk Reduction through forest conservation and management

While flood countermeasures are done by building dams and improving river banks the world over, a method of flood control which takes advantage of the water-retaining capacity of forests is also under discussion by experts, and researchers. Forests have certain functions like "retaining water temporarily and draining it into rivers gradually" and "retaining water and returning it to the atmosphere by evaporation instead of draining it into rivers", that will bring about flood alleviation.

Forest conservation and maintenance by themselves are not sufficient for flood control, however, we should be aware that there are also these alternative approaches to flood control by forest conservation and maintenance, and not only development that causes environmental destruction. This concept should be recognized as it is closely related to the Integrated Water Resources Management recommended at the Water

Summit. the cooperation of each countries is necessary to clarify the relation between water-retaining capacity of forests and flood alleviation.

Furthermore, implementation of forest conservation and maintenance will contribute not only to flood control, but also be a remedy for climate change that is triggering water disasters extensively. And, the forest is important because of the recovery of the culture and the community.

Recommendation for Priority Theme C: Water for Development and Ecosystems

Water is right for all being and it is common property. Water rights should not be determined by market mechanism, which would lead degradation of environment and increase poverty. Based on this concern, we submit following statement.

1) Due to water management systems which ignored bio-diversity (such as construction of dams and river works), caused devastated damage to ecosystems. Therefore, when water related project is implemented, consideration to an environmental and social impact should be put on first priority. Environmental and social assessment should be conducted adequately in early stage, which should include "no project option".
2) Environmental and social assessment should be conducted in order to analyze causes of impacts and specific measures to restore damages should be taken, even if a concerned project is completed. Process to review past failure should be implemented in order to create a mechanism in order to avoid similar failure.
3) Large-scale irrigation system has caused a serious environmental damage such as salination of land, which should be re-examined. Water management systems that are controllable and repairable by local communities should be established.
4) Due to the development which ignored diverse environmental and social function of river, watershed ecosystems, unique local water management and river basin culture have deteriorated. Not only the ecosystems of river, but also the traditional culture and usage of water by local residence should be respected.
5) Asia Pacific Regions which has many trans-boundary rivers need integrated river management. However, basin managements should be led by local communities. Governments and regulatory institutes should assure transparency and participation of communities.
6) Local communities' participation should not be just superficial and autonomous basin management has to be secured. Minority opinion has to be respected and reflected, and local residence and government should have equal claim on basin management.

7) Research proved hydro-electric dam, especially a large-scale ones produce methane gas and thus contribute to the global warming. Therefore, a large-scale hydro-electric dam should not be included as project targeted for Clean Development Mechanism (CDM).
8) Modern agriculture such as "green revolution" caused serious environmental impacts by using massive inputs of chemical fertilizers and pesticides. Based on past failure, sustainable agriculture should be promoted.
9) Those who polluted should be responsible and oblige for water purification.
10) Livelihood-oriented activities such as forestry, agriculture and fishery, and ecosystem are interconnected to sustainability. Thus, integrated watershed management including head water to ocean is necessary. Policies giving an emphasis on basin culture and basin ecosystem are required for a sustainable river basin management.
11) Substantial investment for investigation on conventional water rights and usage is needed.

ANNEX 6

Messages from Open Event Organizers to the Participants of the 1st Asia-Pacific Water Summit

Event Title	The Seminar on the Development of Appropriate Technology in 2007
Date & Time	29th November 2007, 17:00～18:00 (Seminar Guidance) 30th November 2007, 8:00～18:00 (Study Tour) 1st December 2007, 9:00～17:30 (Seminar) 2nd December 2007, 9:00～10:00 (Closing Ceremony)
Venue	Multipurpose hall of Oita-ken new life plaza in Beppu city
Organizer	The Japanese Institute of Irrigation and Drainage

Message from Open Event
Now, while water use in entire world is expected to continue to increase, water for agricultural use, which accounts for approximately 70% of the total water use, presents problems such as lowering of irrigation efficiency due to inadequate on-farm irrigation development and inappropriate operation and maintenance of irrigation facilities in developing countries in Monsoon Asia. In order to promote efficient use of irrigation water, it is necessary to develop on-farm irrigation facilities at low cost. At that time, it is also necessary to sustain functioning of developed on-farm irrigation facilities. For them, it is necessary that the farmers should get ownership of irrigation facilities and manage them by themselves. For that reason, it is important to promote farmers' participation from the planning stage of developing on-farm irrigation facilities and establishment and strengthening of farmers' water management organization as well. Each country in Monsoon Asia is making efforts to overcome of the above mentioned problems. For promoting further efforts, it is efficient that stakeholders in this region gather, share information and discuss on this issue each other. As part of international cooperation programs, the Japanese Institute of Irrigation and Drainage (JIID) has sponsored the Seminar on the Development of Appropriate Technology since 1988 with the support of the Japanese Ministry of Agriculture, Forestry and Fisheries (MAFF). For contributing to agricultural rural development of developing countries, the purpose of this seminar is spreading knowledge in relational countries by exchanging opinions and information among the government officials, engineers and representatives of farmers' water management organizations and searching appropriate measures. This year's seminar was held at Beppu city, Oita prefecture in Japan in time to 1st Asia- Pacific Water Summit. This seminar's theme was "On-farm Irrigation Development and Participatory Irrigation Management" like last year's. In particular, the main issue was Joint Water Management (JWM) by the government public agency and the farmers' water management organization. As its case study, we conducted the study tour to the Ryouchiku Plain Irrigation Region where the JWM was performed sharing the roles between Japan Water Agency and

Ryouchiku Land Improvement District on 30th Nov. At the Seminar on 1st Dec, JIID explained the summary of "Guidelines for On-farm Irrigation Development and Management in Monsoon Asian Countries" which was recently compiled. After that, there were presentations of Japanese Land Improvement System, situations of participants' countries, etc. And we discussed and exchanged opinions based on these presentations.

Participants from overseas countries were Philippines, Viet Nam, Thailand and Indonesia, and observers were Malaysia, Myanmar and Pakistan.

We confirmed the followings as fruit of the seminar.

1. We got understanding about the example of Joint Water Management in Japan by visiting the Ryouchiku Plain Irrigation Region.
2. We will deepen the discussion referring to above Guidelines that were distributed by JIID.
3. As our common understanding, water management of irrigation system can not be well performed by only government public agency or only farmers. It is big issue for discussion that "How the JWM should be conducted by what kind of role sharing?", and this has to argue continuously.
4. We have a big outcome that participants got mutual deep understanding by each others. And it will be necessary for continuous discussion and exchange of information.

APPEAL FOR SMALL-SCALE HYDROPOWER DEVELOPMENT IN JAPAN AND THE ASIA-PACIFIC REGION

December 1st, 2007

symposium for micro-hydropower recovery

-For the construction of "Hydro-net" power stations utilizing irrigation channels-

The Asia-Pacific Region, including the monsoon area, is generally endowed with a rich water supply; and at the same time, it has also had to cope with the destructive power of water. Therefore, people living in the Region have made concentrated efforts to develop technologies and establish appropriate social systems with the aim of alleviating damage from water and maximizing its benefits. The great numbers of artificial ponds, canals, dams, weirs, etc. that now exist are the result of such endeavors. Their construction has been closely connected to the local community with its social institutions, customs and traditions.

Thus, we have a long history of dealing with water and finding a harmonious balance with it. Now we are in a position to make use of water as an energy source ***to the fullest extent***. There is no denying that we have utilized water for power generation on a large and medium scale. The problem is, however, that we have not done as much with small-scale generation, in spite of the fact that there is a huge untapped potential for the latter and social benefits to be derived from it. Consequently, the use of small-scale hydropower is still very limited in the Region at present. We need to change this situation as soon as possible. Doing so would be one of the surest steps in overcoming global warming and paving the way for the sustainable development of our society.

The use of small-scale hydropower has several advantages, such as:

1. The environmental burden is very limited compared with large- or medium-scale hydropower development.
2. A more stable power supply is available than with other kinds of natural energy such as those derived from wind, solar power, etc.
3. It will be able to greatly contribute to the electric power supply of communities as a key local energy resource.
4. It will be useful for electrification in remote and isolated areas, which will improve the quality of life in the areas.
5. It provides economic and social benefits in its catchment area.

In Japan, large- or medium-scale hydropower development has already progressed considerably. On the other hand, the use of small-scale hydropower remains at an almost negligible level at present, despite the fact that there are countless promising places provided by the favorable topography and the extensive irrigation canals.

The development of small-scale hydropower needs different approaches from those used for larger projects. That is to say, institutional and social restrictions for small-scale hydropower development should be removed, and suitable socio-economic systems should be established so that we can start the projects easily and make them economically feasible. Technical support is also needed, including: feasibility studies, consolidation of venture units, training of field engineers, development of equipment and machinery for small-scale hydropower generation, and provision of capital funding. We strongly urge that a social and political consensus be formed to push forward these projects in the various countries of the Region.

In the Asia-Pacific Region, the economies are growing rapidly, while the emission of carbon dioxide is increasing dramatically. It is also reported that the low-lying countries and islands of the Region are exposed to the threat of submergence. In light of this serious situation, we propose that the following measures be taken immediately by the governments, organizations, and institutions concerned.

- Surveys and/or estimates on the potential of small-scale hydropower in each country
- Planning and execution of training for field engineers
- Compilation and publication of guidebooks for starting small hydropower ventures
- Mitigation or abolition of regulations and legislative measures that may inhibit small-scale hydropower development
- Provision of the social institutions to encourage ventures for small-scale hydropower development
- Cooperation among the central government and local governments for the establishment of frameworks for the promotion of small-scale hydropower development
- Establishment of an organization in the Region to promote cooperation and information exchange among the countries

This appeal is supported by:

- The environment conference for citizens of Hita;
- Renewable Energy Promoting People's Forum West Japan;
- HIMUKA-OHISAMA-KYOWAKOKU;
- Group for Utilization of Micro Hydropower Energy in Murasaki River;
- The Society for the Study of Literature about Water, Rivers, Streams, Lakes and Ponds;
- The Center of Network about Water Environment in Hita;
- Green Energy law Network;
- TAO communications&ecovillege project
- Japan Center for a Sustainable Environment and Society;
- Kiko Network.

1st Asia Pacific Water Summit Open Event

Event Title: Beautiful Forests, Beautiful Water
Organizer: Japan Forest Conservation and Flood Control Association
Format: Symposium
Date: Saturday, December 1, 2007
Time: 14:30 - 17:40
Venue: B-con Plaza

1. The Concept and Purpose of the Symposium

The title of this event, "Beautiful Forest, Beautiful Water," embodied the basic concept of this event that growing and conserving healthy, beautiful forests brings our society safe and good water resources. The purpose of the event was to disseminate the concept both domestically and internationally by presenting recent research results and introducing case studies of good practices.

2. Program

14:30 Opening of the Symposium

14:30-14:40 Opening Addresses
Mr. Shouji Fukuda, *Japan Forest Conservation and Flood Control Association*
Mr. Hisao Harihara, Guest, *Director General, Private Forest Department, Forestry Agency, Ministry of Agriculture, Forestry and Fisheries*

14:40-15:20 Keynote Presentation
"Impacts of Degradation of Planted Forests on Runoff and Outflow of Soil"
Dr. Yuichi Onda, Associate Professor, *Department of Integrative Environmental Sciences, Graduate School of Life and Environmental Sciences, University of Tsukuba*

15:20-16:55 Case Studies
"Revitalizing the Ocean and the Earth—To Protect the Headwater Areas of Chikugo River"
Mr. Yuichi Mae, *Nakatsue Village Earth Foundation*
"Diagnosing Forests—Watershed Conservation with Public Participation"
Mr. Kenji Niwa, *Yahagi River Forest Volunteers Association*
"Forest Environment Tax—A New Forest Management Movement"
Mr. Takamasa Kondo, *Department of Agriculture, Forestry and Fisheries, Oita Prefecture*
"Overseas Cooperation—The Forest Management and Community Support Program in Lao P.D.R." (JICA Technical Cooperation)
Mr. Kazuhiro Goseki, *Ministry of Agriculture, Forestry and Fisheries*

16:55-17:35 Special Presentation
"A New Forest Management for Water Conservation"

Dr. Takehiko Ohta, Professor, *Department of Forest Science, Tokyo University of Agriculture*

17:35-17:40 Closing Address

Mr. Shouji Fukuda, *Japan Forest Conservation and Flood Control Association*

17:40 Closing of the Symposium

3. Outline

- About 150 people with various backgrounds attended the symposium.
- Representing the organizer, Mr. Shouji Fukuda declared the opening of the symposium and thanked the participants for their attendance.
- The special guest, Mr. Hisao Harihara of Forestry Agency, saluted the people of Oita and the symposium organizer for their efforts in preparing for the 1st APWS and the symposium respectively.
- Dr. Yuichi Onda, associate professor of Tsukuba University and a leading forest hydrologist, gave the keynote presentation in which he introduced his up-to-date research results. From his field studies and analyses, Dr. Onda suggested that high-ratio thinning is needed to recover the public functions of degraded planted forests.
- Four representatives from various organizations gave case studies about their activities to conserve forests and nurture water resources.
- Mr. Yuichi Mae presented a case of *Nakatsue*, a mountain village in Oita Prefecture where villagers, downstream city residents, and various organizations such as municipalities, corporations, schools, and NPOs gather regularly to grow forests for water and fishery resources. The movement has gained momentum with the increase of the number of participants.
- Mr. Kenji Niwa, an acknowledged leader of forest volunteer groups, advocated a new method of forest inventory which was specially devised so that amateur volunteers can enjoyably take part in the field work of forest inventory and yet contribute to professional forest analyses for better forest management.
- Mr. Takamasa Kondo of Oita Prefecture described the background and the system of a special tax called forest environment tax that Oita Pref. recently introduced. In consultation with citizens, the tax revenue is expended on projects related to forests such as public works of forest conservation, forest products promotion, and forest education for children and adults.
- Mr. Kazuhiro Goseki presented his experience as a Japan International Corporation Agency (JICA) professional dedicated to the forest conservation and afforestation project in Vientiane Province, Lao P.D.R. The community-based skill transfer approach of the project established participatory forest management and income generation methods for villagers.
- Dr. Takehiko Ohta elaborated the headwater conservation function of forests in the light of human's forest utilization and management history. Dr. Ohta propounded that forest management from now on should well consider the characteristics of multifunctional roles of forests and plan forest conservation according to the expected main function of each forest.
- On closing, Mr. Fukuda expressed his appreciation to all the people involved.

*This event is partly sponsored by the National Fund for Forest Greenery and Water of Japan.

Chubu Declaration

Today, in the twenty-first century, the environment of Earth, the "water planet," has been brought to a crucial crossroads. The threats to the environment often manifest themselves in water-related phenomena, such as droughts, floods, disappearing glaciers, drying up of lakes, and water pollution, which can be expressed as dangers to the "Life of Water."

Japan, and particularly the Chubu region, has developed through manufacturing activities. Chubu is an area that has been subject to severe geographical and meteorological conditions and various water-related hazards. Floods and tidal surges have claimed the lives of many people living in low-lying areas. It has suffered through debilitating water shortages, and groundwater pumping has led to ground subsidence and water contamination.

The people of the Chubu region have overcome these dangers by building up social infrastructure, learning to share scarce water resources, protecting river sources, and maintaining water quality. Chubu-area manufacturers have not only engendered world-leading energy-saving technologies but have also applied traditional technologies to achieve the coexistence of human activity with a healthy water cycle.

These experiences have been passed down to the present generation, and both leading-edge and traditional technologies have been ingeniously enhanced so they may be easily used by anyone. The history and technologies of Chubu manufacturers have evolved into cultural and intellectual assets, enabling the development of cutting-edge modern industries in the region that coexist side by side with such primary industries as agriculture, forestry, and fisheries.

This symposium has been organized in the light of current environmental needs and the historical achievements of the Chubu region and is grounded in the awareness that building healthy water cycles in regions around the world can help bring about sustainable global development. Based on the presentations and opinions expressed by the participants of this symposium, all participants join in endorsing the following declaration.

Declaration

Japan's development is supported by the resources and people of countries around the world. And as a member of the international community, Japan is responsible for helping protect the global environment.

Chubu announced its intentions to contribute to the world's environment by hosting the 2005 World Exposition, Aichi, Japan.

The responsibility of the Chubu region, as the center of Japan's manufacturing activities, is to continue conveying its experiences and technologies to the world. Such initiatives will attract the latest information to the Chubu area, and the accumulation and interaction of global information flows in Chubu will lead to the birth of new technologies and knowledge that can contribute to the resolution of the world's environmental problems.

In the twenty-first century, the Chubu region will continue to be a source of technology and wisdom for the world. The diverse range of technologies accumulated in the region will enable it to respond quickly to environmental problems around the world and contribute to their resolution.

The Chubu region will also continue to closely monitor environmental changes both locally and globally and to apply this knowledge to the achievement of healthy water cycles. We believe that this will also lead to sustainable future development for the Chubu region itself.

For this purpose, we hereby boldly declare that the region's industry, academia, civil society, and government will develop a stronger awareness of their respective responsibilities and proactively pursue initiatives. Working closely with one another, we will also seek to actively share information and to cooperate with people around the world.

The Water Cycle

December 1, 2007

The Participants of the "Life of Water" International Symposium

Recommendation of IFNet
"Towards Flood Disaster Reduction"

Conveners : International Flood Network (IF-Net)
Date & Venue : 1 Dec. 2007 from 13:30 to 18:00 Hours
Beppu city Social Welfare Hall

1. Approaches from Whole Society

Since flood disasters are increasing as a result of various socio-economic factors, it's difficult to solve the problems without suitable approach to the primary causes of disorderly urbanization, poverty, recent climatic change in the world and so on. Therefore under the suitable share of roles, approaches from whole society to raise resilience against floods are requisite.

2. Optimum Combination of Structural and Non-Structural Measures

In response to individual regional conditions, the optimum combination of Non-Structural and cost-beneficial Structural measures should be steadily developed with full support from the local communities so that they could have a maximum effect on disaster reduction.

3. The Measures towards Early Warning Leading to People's Evacuation

The Measures towards Early Warning Leading to People's Evacuation, which is one of Non-Structural Measures, should be respected more and pressed for further reforms to be made.

Because prevention of human losses linked to water-induced disasters such as floods, tsunamis and debris flows is possible if adequate actions based on correct judgment of early warning are taken before the likely disaster materializes. This is a sharp contrast to the case of earthquakes which are less predictable and occur momentarily.

Given that even those areas which are far from regions having heavy rainfall, domestic or sometimes cross-border, would suffer from flooding of big rivers, early warning of which is transmitted extensively bears a great significance. Such cases occur often, however, where inadequate information transmission failed to convey important disaster information properly. For reducing human loss in a short time, it will be efficient to improve and reinforce the current information transmission systems, leaving no zones uninformed.

Early warning doesn't necessarily lead to prompt evacuation in those areas without any experiences of past disasters or with relief generated by completed structural measures. The way of providing information leading to prompt evacuation in dangerous areas should be examined.

Network of Asian River Basin Organizations (NARBO)
– Symposium on Catalyzing IWRM investment in the Asia-Pacific Region–
Report to the 1st Asia-Pacific Water Summit

1 December 2007, Beppu City Social Welfare Hall

Participants from 11 countries met on the eve of the 1st Asia Pacific Water Summit to discuss key challenges for integrated water resources management (IWRM) in the region, and future directions for NARBO to assist governments and river basin organizations (RBO). The Symposium was convened by Japan Water Agency, Asian Development Bank and ADB Institute as NARBO Secretariat, in collaboration with the Ministry of Land, Infrastructure and Transport and the Infrastructure Development Institute of Japan. Key points discussed were as follows.

Recognizing that:

1. The eastern and southern parts of the Asian Continent and the northern parts of Indonesia and Australia are influenced by a monsoonal climate under which distinct patterns of rainy season / dry season emerge in the region, and also precipitation varies substantially on a monthly basis. Thus, the Asia-Pacific Region encounters either problem of too much water or too little water by seasonal changes of the climate. This tendency of "too much water or too little water" becomes more significant due to the global climate change.
2. Generally speaking in the Asia-Pacific Region, the upper reaches of river basins which are steep and located at near- plate boundary zones are geologically vulnerable to volcanic eruptions and earthquakes. On the other hand, the lower reaches of river basins are formed of flood-prone alluvial plains. Once heavy rains hit the river basin areas, flood water flows quickly down to the sea causing sediment disasters such as erosion, landslides, debris flow and so on in the upstream areas and lethal impacts on socio-economic activities, specifically human lives and assets in the downstream areas.
3. Water demand is projected to increase with continued economic development in the Asia-Pacific Region. Disaster risk increases concurrently as more people and assets are accumulating in the flood-prone alluvial plains than ever before. The region's rich natural environments are already degraded and under further pressure. Water resources and ecosystems need to be protected to ensure sustainable development to future generations. Governments need to improve their water infrastructures and strengthen their institutional capacity to address these challenges through effective water resources development and management.

Appreciating that NARBO's 56 member organizations are starting to make an effective contribution to the introduction of IWRM in river basins across the Asia-Pacific region since the network was established in 2004, the participants adopted the following directions for NARBO activities in the coming years:

1. NARBO will facilitate the implementation of IWRM in the Asia-Pacific Region taking account of its characteristic features, including the occurrence of water-related disasters arising from floods and droughts, the continuing increase of water demand due to economic development and urbanization, and widespread deterioration of water quality and environment in river basins. To help RBOs develop locally appropriate solutions in river basins, NARBO will elaborate the IWRM approach in practical guidelines that reflect the rich and diverse experience across the region. The guidelines will be prepared in collaboration with member RBOs and knowledge partners, in time for presentation to the 5th World Water Forum in 2009 in Turkey.
2. Drawing on its first five years of experience, NARBO will also prepare a long-term plan for introduction of IWRM in river basins across the region, to help governments and RBOs achieve the MDGs and respond to the unprecedented transformation facing the region in the management of water resources, including the increased variability and risks brought by climate change.
3. NARBO will continue to support governments and RBOs in improving the enabling environment for IWRM, including policies, legislation, institutional development, and public awareness and participation. NARBO will also continue to facilitate the sharing of knowledge and experience in the development and management of water resources, in priority topics such as water rights, disaster management, water quality management, and climate change adaptation. NARBO's performance benchmarking and peer review service will be expanded, and NARBO will pilot an advisory service for RBOs in the preparation of integrated long-term investment programs in river basins.

The 4th International Workshop on Coastal Disaster Prevention

Future Disaster Management against Tsunami and Storm Surge in Asia-Pacific Region

Open Event of the 1st Asia-Pacific Water Summit

December 1 – 2, 2007
Yokohama Symposia

Devastating coastal disasters have occurred in the world recently such as the Indian Ocean Tsunami in 2004, the storm surge generated by Hurricane Katrina in 2005, the Java Earthquake Tsunami in 2006, and the storm surge by Cyclone Sidr most recently. It is possible that tsunamis and storm surges would also cause more serious disasters in the future, considering sea level rise due to Global Warming.

We have learned a great deal from the experiences of recent disasters including rescue and recovery activities after the disasters. Intensive studies have been conducted to develop more effective and economically feasible countermeasures in the world, especially in the Asia and Pacific region.

This workshop is the fourth meeting after the first one in 2004 just after the Indian Ocean Tsunami disaster. At the first workshop, researchers, engineers and government officials gathered from Asia and Pacific region and discussed future technologies and management systems to reduce tsunami and storm surge disasters especially in the Asia-Pacific region considering the recent development after the Indian Ocean Tsunami disaster.

Sessions:

Technical Session 1 – Countermeasure against Storm Surge Disaster
Technical Session 2 – Advanced Technologies of Tsunami and Storm Surge Disaster Prevention/Reduction
Technical Session 3 – Information for Disaster Prevention
Technical Session 4 – Lessons from Disasters
Technical Session 5 – Future Disaster Prevention Strategy

Panel Discussion: Future Disaster Management for Tsunami and Storm Surge in Asia-Pacific Region

Organizers:

- Port and Airport Research Institute (PARI), JAPAN
- Coastal Development Institute of Technology (CDIT), JAPAN
- Ministry of Land, Infrastructure and Transport (MLIT), JAPAN

Supports:

- Port and Harbor Bureau, City of Yokohama, JAPAN
- Japanese Section of International Navigation Association (PIANC)
- Panel on Wind and Seismic Effects, U.S.-Japan Cooperative Program in Natural Resources (UJNR)

1st Asia-Pacific Water Summit Beppu Japan 2007

PARI

CDIT

Ministry of Land, Infrastructure and Transport

1993 Hokkaido Nansei-oki Earthquake Tsunami, Okushiri Island

Summary of the Workshop

1. We will need to develop holistic coastal disaster prevention and mitigation measures using integrated technology:
   - More reliable early warning systems
   - More accurate and detailed disaster prediction
   - Possible worst case prediction
   - Effective and easy-access evacuation
   - Effective structural measures to reduce economic losses

2. We will need to encourage disaster preparedness and participatory regional disaster prevention through integrated technology:
   - Creation and transmission of clear disaster information for public
   - Disaster preparedness with due considerations of rescue, restoration and reconstruction based on the disaster scenarios including the worst case
   - Disaster preparedness as a part of coastal management and community planning and development

3. We will need to enhance cooperation among the Asia-Pacific region in order to exchange scientific and technical information and to foster joint research.

2004 Indian Ocean Tsunami, Sumatra Island

2004 Indian Ocean Tsunami, Sri Lanka

2007 Solomon Islands Earthquake Tsunami Ghizo Island

Asianchildren water summit

アジア子ども水サミット　2007.12.1(sat)~2(sun)

子ども水宣言 Children Water Declaration

「理想の水辺とは、・・・・」　・空気がきれいなこと　・源流がはげ山ではない　・水質がよい
・きれいな安全な川　・楽しめる川遊べる川　・きれいな花が咲いている
・いろんな生き物がいる　・風、太陽の力を生かせる川　・周りの環境がきれいな川

①川のまわりの環境がきれいで豊かなこと
②ゴミがなくて水質が良いこと
③魚や水鳥や木や花などの生き物がたくさんいること(天然記念物も！！)
④風や太陽などの自然エネルギーを利用できること
⑤災害のない安全な川
⑥僕たち私たちが川の中で楽しく遊べること

守江湾・八坂川河口で
干潟探検しました。

身近な水辺について
考えました
持ち寄った絵を使って説明しあいました

○よいところ、好きなところ
生きものがいっぱい、水がキレイ！、遊べる！
人に会える！、川のまわりがキレイ

×わるいところ、きらいなところ　　あれてきた　汚れてきた　ゴミがある！もっと便利になるとよい！

Japanese and Chinese schoolchildren gathered with the paintings of water in their daily scenes. The pictures shows the reality of water connected with our daily lives.

The Japanese and Chinese schoolchildren has got to know each other and collaborated in drawing various ideal scenes with water. They are impressing us with the preciousness of water, which grows and cares all forms of life.

Water Message Committee
From The Central Japan
中部からの「水」メッセージ委員会

事務局 ： 国土交通省 中部地方整備局 庄内川河川事務所 地域連携課

（株）INAX・王子製紙（株）春日井工場・日本ガイシ（株）・（株）山田組・NPO法人Waterscape・愛知県・名古屋市・春日井市・国土交通省庄内川河川事務所・(財)河川環境管理財団名古屋事務所

Event Name: 1st Youth Asia-Pacific Water Summit (4th College Student Activity Exchange)
Event Organizer: 1st Youth Asia-Pacific Water Summit Executive Committee
Date: December 1 (Sat) and 2 (Sun)
Contact (Mr. SAKAMOTO): joint_of_college@yahoo.co.jp

December 2, 2007

"Asia-Pacific Youth Water Declaration"

We the youth hereby proclaim to establish the Youth Disaster Reconstruction Volunteer Network with the rising awareness of disaster threats and in aim for the earliest possible reconstruction of the afflicted areas.

We will exert our youth power to bring back the smiles on the faces of the victims and will actualize prompt support to revitalize the afflicted areas. When doing so, we will also be fully aware of the possibility of secondary disasters and will voluntarily hurry over to the afflicted areas. We will precisely identify the emergency information of such areas and will immediately spread information extensively by utilizing the WEB, etc. Moreover, we will learn from the knowledge behind examples and disaster prevention of damages caused by wind or flood and will build up a network with other youth through continual exchange.

We the youth will then, by improving our skills, strive to demonstrate leadership. Additionally, we will send out information pertaining to methods of disaster support and will provide more opportunities for information exchange through enjoyable interchange with those in the community by participating in local activities. In case of any disaster, we will continue deepening our relationship with the local people and value human bonds.

We will also create a youth action map that specifically identifies the relationship of various activities each youth is involved in the respective regions.

In respect to such activities, we will communicate our know-how and thoughts to junior high and high school students. Furthermore, so that we may cooperate even more quickly with one another in emergency situations, we will provide opportunities for exchange on a regular basis where college students from various regions can get to know each other and will work hard to enhance the sharing of information and networking.

In all, we the youth promise to do our utmost to ensure that our wishes will be actualized.

English

Abe River:the Crystal Waters that Gave Rise to a City for the Ages

On November 7th, 2007, Shizuoka City, with the consensus of its citizens, inaugurated the Shizuoka Environmental Promotion Council in a declaration of its determination to contribute to discovering water-related problems and solving such problems in the Asia-Pacific region.

Abe River

Shizuoka, the City of Crystal Water, is found in the centre of the Japanese archipelago, surrounded on one side by the mountains of the Southern Alps, and on the other by Suruga Bay, which at over two kilometres below sea level, makes it the deepest in all Japan. In the midst of this amazing environmental contrast lies a city and the beautiful river which has supported it through the millennia.

Japan is well known for its geological fragility, and the Shizuoka area is no exception. A fast flowing river, the Abe carried down collapsed sand and earth through the ages, forming the fan-like Shizuoka Plain and the Miho Peninsula, which are the foundations of the Shizuoka City of today.

The Toro Ruins of the 2nd and 3rd centuries AD speak of the rice-cultivating community established here, and in the 8th century, the inhabitants set up this place as the capital of the domain known as Suruga-no-Kuni, providing the ground work for all future development.

As Japan moved through periods of great transition, this land came to be known as Sumpu, and held great importance in the workings of the burgeoning nation. Tokugawa Ieyasu, the man who unified Japan and brought stability to the country, spent his last years in Sumpu, and during his reign directed a great project to divert the course of the Abe River, to make it more usable as a water supply for the town. And as the Shizuoka City area has grown to include all of the river area, from mountain to farmland to ocean, its people aim to continue the tradition of sustainable urban development.

The water purified by the vast forests aids in the growth of local agricultural products, especially green tea and wasabi, and sand and gravel layers have filtered and preserved the abundant underflows in springs and underground waters that have enriched this area's inhabitants' lives for centuries. Suruga Bay, supported by plant material brought by Abe River, plays home to a wide variety of underwater life, and is the source of great bounty, including the popular sakura shrimp and shirasu baby sardines.

In olden times, however, in heavy rain the Abe would turn violent, and place the people in great danger of flooding and landslides. To protect themselves from these threats, the people of Shizuoka have undertaken a long series of flood control projects, starting with the construction of dikes built by engineers from across Japan under the guidance of Tokugawa Ieyasu, and which continue to the present day.

The Shizuoka municipal water management policies have been designed to preserve the water cycle and ecology of the Abe River, and continue the process of sustainable development. While following national and prefectural water management guidelines, the city government and citizens have been working together without outside involvement to maintain the health and character of this valuable resource.

Specifically, the city government has enacted the Crystal Waters Ordinance, and concluded conservation agreements with local agricultural and forestry cooperatives that directly enjoy the benefits of the Abe River, which are meant to preserve our forests' natural water cultivation system and the health of the river. Additionally, the government continues to maintain and improve the city sewer infrastructure and upriver water purification facilities. It also provides support to citizen-based cleaning activities and an Adopt-a-Riverbank program. City Hall has also implemented sustainable project financing, known as the Forest Environment Fund, for forestry conservation and management activities.

These actions are very important parts of a water management policy that is meant to protect the natural character and condition of the river system while encouraging citizen-centered preservation activity. This continues the tradition of fostering a heightened sense of sustainable water cycle management, of learning to subdue the threats of the Abe River while taking great care in its use. This tradition of careful coexistence began with the founding of the Toro settlement millennia ago, and with care, will continue into the future.

Even though Shizuoka has served as an industrial, economic, and residential focal point for centuries, the Abe River running through it all has remained as the purest river in all Japan; a gift passed on through the generations. This in itself may be a small miracle, not just in Japan, but perhaps even in the entire Asia-Pacific region.

It is our fervent hope that, as we introduce the water management policies of the Abe River as an example of nearly two thousand years of continuous urban development, we can use the 1st Asia-Pacific Water Summit as an opportunity to build and strengthen ties with the many participating governments and local authorities, and to contribute to the discovery of solutions to the water problems we and our neighbours face.

Zenkichi Kojima, Chairman of the Shizuoka Environmental Promotion Council and Mayor of Shizuoka City
November 7, 2007

The Crystal Waters of the Abe River Exhibit
Organizers : Shizuoka City Dates : 1-4 December 2007

Love H_2O Declaration

December 6, 2007
Team Love H_2O

W We cannot live for even one day without water.
Yet, we do not give it the credit it deserves in everyday life.
"Water can be found anywhere."
"Water won't run out, like oil will."
··· Is that really true?

A All of us must take note. All of us must face the fact that one in every five people around the world has difficulty accessing safe drinking water.
All of us must become aware that, in order to bake that loaf of bread you eat, water several thousand times the amount of flour that goes into that bread is being consumed somewhere in the world.
We must learn how global warming is gradually distorting the circulation of this precious water.

T Take note of the need to protect water and protect lives supported by water. Take note of the need to protect the only planet with water in the universe, the mother planet that continues to generate various forms of life. We must start taking small yet steady steps to do what should be obvious.

E Explore the notion of "eternity" and cast your thoughts toward the joys and sorrows of future generations. Transcend "national borders" and bring our hearts together as fellow inhabitants of this beautiful planet of water "Earth", to examine the future of water and the Earth.

R Remember; the only rule is to keep loving "water", the source of all life.

Today, from the Japan Pavilion, Expo Zaragoza 2008, we are taking our first step into the future.

Public Relations Booth for EXPO Zaragoza 2008
Japan External Trade Organization(JETRO) From Dec. 1 to Dec.4

The 1st Asia-Pacific Water Summit – Official Open Event
JTUC-RENGO & PSI Symposium—Declaration on World Water Challenges and Japanese Standpoints

On December 2, 2007, JTUC-RENGO (Japanese Trade Union Confederation) and PSI (Public Services International) held a symposium entitled "World Water Challenges and Japanese Standpoints" at the Suginoi Palace in Beppu-shi, Oita Prefecture, as an official open event of the 1st Asia-Pacific Water Summit, with the participation of 200 people.

The first section of the symposium, organized by JTUC-RENGO, featured a report from the Summit steering committee; a lecture "On Japan's Water Issues"; a report on the present state of waterworks issues, and a policy proposal from JTUC-RENGO. The participants forged a common recognition of the present situation and challenges involving water in Japan, achieved a deeper understanding of the significance of a Basic Water Law, and noted the importance of working together toward its legislation.

In the second section, sponsored by PSI, we shared PSI's keynote proposal "Strengthen Public Water Systems for Quality Water Services"; a presentation by by PSIRU (Public Services International Research Unit) "Public Public Partnership in the Water Sector"; reports on overseas cases by AGWWAS (Philippine Alliance of Government Workers in the Water Sector) on "Water Challenges for People – Centered Model" and by KPTU (Korean Federation of Public Service & Transportation Workers' Union) on the "Problems of Water Privatization and Responses in Korea," and a presentation on "Public Water Management in Japan" by NPO AM-Net (Advocacy and Monitoring Network on Sustainable Development). Through these presentations, we achieved deeper understanding of the current situation and possible solutions to global water issues focusing on the Asia Pacific region.

JTUC-RENGO and PSI, as organizers of this official open event of the 1st Asia-Pacific Water Summit, adopted the following declaration for the solution of the Japanese and global water problems:

1. All the people of the world should have equal and continuous access to safe drinking water and sanitation services. In achieving this, public waterworks have a very important role to play and must contribute to the achievement of the UN Millennium Development Goals (MDGs).

2. Water is a basic human right, not a commodity. All people should have access to high-quality public services, and waterworks should be provided as a public service under public management.

3. Governments, the ADB and other IFIs, and donors, should, in the formation of policies aimed at the solution of global water problems, guarantee the participation of all stakeholders in the decision-making process. We must build a new framework to promote dialogue with civil society, in a way that is not top down, but that is open to users, residents, trade unions, and others, under the principles of Transparency, Accountability and Participation (TAP). We encourage public-public partnerships and labour-management cooperation to improve public waterworks.

4. Japanese water supply entities have endeavored to supply "Clean, ample, and affordable water" based on the Water Supply Law, a piece of legislation guaranteeing the right to existence as prescribed in Article 25 of the Constitution. Under the principle of municipal management, the system has earned a solid reputation based on the stability of quality, quantity, and management. While overcoming problems that exist, we should continue to "provide safe water that is fit to drink" in the 21st century in consideration of the "Waterworks Vision," a goal set by the Ministry of Health, Labour and Welfare.

5. The provision of water and sanitation services in a safe and qualitatively and quantitatively stable way is the basis for peace, welfare, and local independence. Comprehensive water cycle management should be carried out to secure safe and stable water. We need to make efforts for the early realization of a Basic Water Law under the philosophy that "Water is common property" and "Water is a basic human right."

6. In Japan, where nearly all people drink tap water, there is a need for the government, municipalities, and residents, as well as enterprises and organizations involved in waterworks, to deepen their partnerships for the public management of water and to use their experiences to contribute, as world front-runners, to national and international cooperation through public-public partnerships. Let us work for the establishment of self sustaining local water supply and sanitation systems based on the principle of "Public Ownership and Public Management."

December 2, 2007
JTUC-RENGO (Japanese Trade Union Confederation)
PSI (Public Services International)

THE OPEN EVENT RELATING TO THEME C "SYMPOSIUM ON RIVER ENVIRONMENT AND WATER RESOURCES MANAGEMENT"

Leading organizations:	**Symposium Committee of River Environment and Water Resources Management**
Time & Date:	**9:00 to 12:00, 2 December 2007**
Venue:	**Social Welfare Hall (Shakai Fukushi Kaikan), Beppu City**

Activities in management of water resources and the river environment in Japan could be advanced attempts and important experiences in the Asia-Pacific region. In order to effectively adopt such experiences to other countries in the region, relevant experts from Japan and the Asia-Pacific region were brought together and discussed on the river environment and water resources management.

RECOMMENDATIONS

- Since water resources development requires time and cost, this should not be adopted as an ad-hoc strategy, aiming at short term gains in water demand, but as a long term national strategy enabling sustainable use of water resources. At the same time, water resources development, which requires dam construction, has various impacts on the river environment. Therefore, in the case dam development is needed, environmental assessment must be appropriately carried out. In water resources development, effective use of stock in existing works should also be considered.
- Considering various impacts of existing works on the river environment, there is a need for establishing a new framework for environmental river flows, which is applicable to Japan with its natural and social conditions. Developing flexible operation of dams could be one of the effective tools in order to put such framework into practice.
- Japan has been implementing nature friendly river management to improve environment of river channels, which aims at balancing flood management, water resources management and the environment. Toward its successful goals, extensive and continuous efforts are required.
- Challenges of the river environment management, including ecosystem management, are to address the river channel as well as river flows and water quality. And management of the river environment with a perspective of river basins is important.
- In the Asia-Pacific region, which has the large population and metropolises with rapid growth rate of the population, water resources development still remains required. However, environmental issues also need to be addressed. Even though lessons learned and experiences gained in Japan in balancing water resources development and the river environment are evaluated as advanced attempts, they could not be directly applied to other Asia-Pacific region because of differences in geology and climate. In order to apply them to other countries, further cooperation and implementation are required among countries in the Asia-Pacific region including Japan.

Leading organization: Symposium Committee of River Environment and Water Resources Management
Partner organizations: Committee on Hydroscience and Hydraulic Engineering; Ecology and Civil Engineering Society; Japan Society of Hydrology and Water Resources; Ministry of Land, Infrastructure and Transport; Japan Water Agency; Water Resources Environment Technology Center; Foundation for Riverfront Improvement and Restoration; and Foundation of River & Watershed Environment Management

Water - Connecting the Human Life and Wild life of Oita, Japan and the Asia-Pacific Region

水でつながる大分、日本、アジア太平洋の生きものと人

December 2nd, Sunday 9:30–16:00

Part I **"Water circulates, generates life and connects people"**
Suggestions for the conservation of the water environment and the use of coastal areas around Nakatsu tideland

Part II **"For the Conservation of Biodiversity "**

Part III **"Protecting and Revitalizing the Water Environment Starting with the Local Community"**

The Proposal of our open event in the 1stAPWS

1.It is needed for multisectoral to deepen mutual understanding to the water environmental conservation.

2.A system of society must be formed to reflect the will of multisectoral in the policies.

With the model of the partnership between Korea and Japan, it is needed to fix the network to promote the collection of marine litter and to restraint producing litter by the cooperation of the governor and citizens.

The conservation of biodiversity will be promoted in and out of Japan.
1. Revaluating of agriculture and popularizing the ecosystem service brought from agriculture, forestry and fisheries
2. The cooperation between primary industry and the activities for conservation

On the basis of our discussion, local activities and local knowledge can contribute both domestically and overseas, as well as debate the issue of whether international activities can contribute to local communities. We also looked into the meaning of local networks based on an international meeting.
1. Find and collect domestic original cases on "Securing Environmental Flows". And position them international context.

We propose to form following domestic and inter Asia-Pacific countries network:
1. Network for Monitoring the Coastal Environment at 60m Deep
2. Island Water Cycle Conservation and Restoration Network

Organizer **Mizubeni Asobu Kai (NPO)**
Oita Biodiversity Conservation Center (NPO)
Partner **Oita pref.**
Dr.Satoquo SEINO Laboratory (The University of Tokyo)

1st Asia-Pacific Water Summit Beppu Japan 2007

International Water Association

JIP-Water 水道国際貢献推進協議会

"Regulation and governance" is arguably a soft discipline but when soundly expressed and delivered, it is a powerful one.

A Communiqué to the First Asia-Pacific Water Summit from the 3rd IWA Workshop "Regulation and Governance in Water Supply" Sunday 2nd December 2007, Beppu, Japan.

If we accept that water is a fundamental human right, we must also accept that it is a core responsibility of governments to ensure that their people have access to safe drinking water.

There are some 1.1 billion people in the world who do not have access to safe drinking water and approximately 60% of these are in the Asia-Pacific region. In many countries drinking water supply facilities have not been properly maintained or managed, or aging facilities have not been renewed. In other cases the necessary infrastructure does not exist or development is surpassing the capacity of existing water supply systems.

Sometimes we do not have adequate funding or human resources to tackle these needs. But even where the people and funds are available we observe that some countries or utilities are successful and others are not. This suggests that our challenges in water supply are not just engineering in nature, they will often rely on the structures and frameworks within which we operate.

At the 3rd IWA Regional Workshop in Beppu, Japan, participants explored various cases of regulation and governance and concluded that soundly expressed and delivered governance is integral to the efficient and effective use of human and financial resources, and in turn, to attain the MDG (Millennium Development Goals)

The meeting recognized that there is no single exemplar of good regulation or governance systems in water: effective systems may vary between countries – this will depend on the prevailing economic, social and governmental systems. There are however several key considerations that are universal:

1. Water supply systems have long lives and long lead times for their implementation. Accordingly it is essential that there is robust long term planning process and capability.
2. Both the quality and quantity of water supply is dependant on water being considered holistically as part of an integrated cycle. We must therefore consider the broader environment from which we draw our water. This requires protection of our aquatic ecosystems through urban planning, agricultural land use and catchment management and sanitation/sewerage treatment as integral aspects of securing drinking water supplies. Further we must recognize that Climate Change will increase the complexity and enormity of our task and will warrant action in adaptation and mitigation.

International Water Association

JIP-Water
水道国際貢献推進協議会

3. Strong policy setting and regulatory structures are the foundations upon which plans will be set and delivered and performance monitored

In many nations there will be a combination of public and private involvement in water supply. Regardless of who delivers services, what is clear is that governments must fulfill their responsibility to set expectations and regulate delivery for the benefit of its people.

Knowledge sharing and capacity development will also be key to achieving the MDG's and the meeting reviewed examples of various practical tools and solutions. Benchmarking and the use of Performance Indicators have the potential to support the regulatory or oversight function and to be important business improvement tool. Both knowledge sharing and capacity development can be significantly assisted through the use of programs such as Water Operator Partnerships which allow utilities to twin with each other on a not-for-profit basis.

The sponsor organizations for the workshop have given their commitment to developing these tools further. Working with governments, this will be achieved through professional associations – such as IWA and JWWA – and adjunct programs.

The 3rd IWA Workshop "Regulations and Governance in Water Supply"
Organizer:
IWA (International Water Association),
JIP-Water (Japan International Partnership for Sustainable Water Management)
Contact:
JIP-Water c/o Japan Water Works Association, Mr. Yoji Matsui (Email:matsui@jwwa.or.jp)

Messages from the Symposium at Beppu, Japan on: "Integrated Approach to Water-Related Disaster Management"

Conveners:	River Bureau, Ministry of Land, Infrastructure and Transport (MLIT), Japan, and International Centre for Water Hazard and Risk Management (ICHARM)
Date & Venue:	2 Dec. 2007 from 13:00 to 18:00 Hours Beppu city Social Welfare Hall

High-level guests from international organizations, academic experts and government officers convened and extensively discussed on the water-related disaster management. The main focus was the anticipated Climate Change. It was agreed that adaptation was as important as mitigation in response to the implications of Climate Change and elaborated on flexible adaptive measures to realize an integrated approach towards water-related disaster management including coping with Climate Change. Water-related Disaster Risk Reduction (DRR) is the main means of adaptation to Climate Change, which is urgent and necessary. Based on the discussion, the following messages are adopted for consideration by the participants of the1st Asia-Pacific Water Summit.

Key Messages

1. In the Asia-Pacific region, even under current geographical, meteorological and socio-economic conditions, the level of Water-related DRR Management is far from sufficient and many vulnerability factors are ever-increasing. As reported by the IPCC, it is now evident that the risks resulting from Climate Change may cause extra burdens such as extreme precipitation and tropical cyclones, more frequent droughts, enlarged risks from storm surges. Countries need to create an enabling environment and empower people, communities and local governments including sharing knowledge, resources and information, raising awareness and ensuring participation in the policy/decision-making processes to cope with threats from Climate Change.

2. Each country should effectively combine both structural and non-structural measures, depending on its specific conditions to reduce the risks from water-related disasters, thus contributing to the promotion of water-related DRR to protect people, property and well-being. While non-structural measures should be more effectively adopted, structural measures will continue to play an important role to improve the protection level from expected risks, for which resources need to be ensured for steady implementation of necessary measures. It is imperative to complement national allocations by focused investments from development banks and development assistance, including ODA.

3. From the viewpoint of protecting human lives, it is necessary to circumvent risks by ensuring timely and smooth evacuation at the time of disasters as well as disaster restoration, which will be provided by measures such as hazard maps and early warning systems. Furthermore, adaptive measures should be taken in a socially acceptable manner, and with sustainable development perspectives in mind. For this purpose, governments should continue to bear the overall responsibility on nation-wide policy planning, resources mobilization and implementation in cooperation with local governments, communities and relevant agencies. As agreed by the Ministers during the "Ministers' Forum on Asia-Pacific Infrastructure Development" on 30 August, 2007 in Beijing, water-related DRR needs to be realized within the context of Integrated Water Resources Management (IWRM) which includes proper management of excessive as well as scarce water. Integration needs to reflect various social, economic and cultural situations in the region, recognizing the importance of internationally adopted frameworks such as Millennium Development Goals and Hyogo Framework for Actions.

4. Water-related DRR needs to be incorporated, with the highest priority, into national development plans with clear goals/targets. Efforts need to be made to establish a standardized procedure for disaster preparedness to monitor and better manage water-related disasters.

5. Quantitative effects of Climate Change on hydro-meteorological conditions should be carefully gathered to formulate adaptive strategies thus enabling actions be taken efficiently on time. The Asia-Pacific countries should share experiences, methodologies, technologies and resources and synergize to comprehensively and properly promote water-related disaster management where willingness for contribution was expressed by Japan, ADB and many others. It is also important to create a knowledge sharing platform (knowledge hub) and to endeavor to foster expertise and enhance institutional capacities based on local needs. The scope of this platform should include scientific research, data acquisition and processing and enhancement of monitoring and assessment capabilities.

Towards Water-related Disaster Risk Reduction in the World

Demonstration of the Flood Forecasting System Utilizing Satellite-based Rainfall Data

Message from the Openevent

Date : December 2nd to 3rd, 13:30-18:00
Venue : B-con Plaza
Organizer : International Center for Water Hazard and Risk Management (ICHARM)
International Flood Network (IF-Net)

Water-related disaster is one of the challenges we should overcome to achieve sustainable development and poverty reduction. Its effect is increasing with population growth, and concentration and increasing value of assets in the flood plain in recent years.

In the countries where river improvements are not sufficient, smooth evacuation from flooding is important for decreasing loss of life and properties. Dissemination of risk by hazard maps, etc. and direction of evacuation by issuing flood forecasts and alerts are necessary. However, in reality, development of a flood warning system in these countries has not advanced properly because of financial difficulty, lack of rainfall data, etc.

ICHARM and IF-Net are developing flood a flood forecasting and warning system using precipitation data from satellites in order for the countries lack of surface precipitation data to establish flood forecast system rapidly and effectively. This system uses the satellite-based rainfall data, and has modeling function, a engine for runoff analysis, a input-output interface and result display function, which enables runoff calculation necessary for issuing flood forecasts and alerts.

In this presentation, we demonstrated the flood forecasting system we developed so far which utilizes the satellite-based rainfall data in addition to the explanation of the development background and its main function.

Also in the countries lack of surface precipitation data, by utilizing this system introduced this time, the flood forecast system is established rapidly and effectively, and it is thought that it leads to mitigation of a water disaster risk.

Moreover, in ICHARM and IF-Net, while continuing development, it is due to raise predictive accuracy and to improve so that it may become the system which suited the needs of the area and.

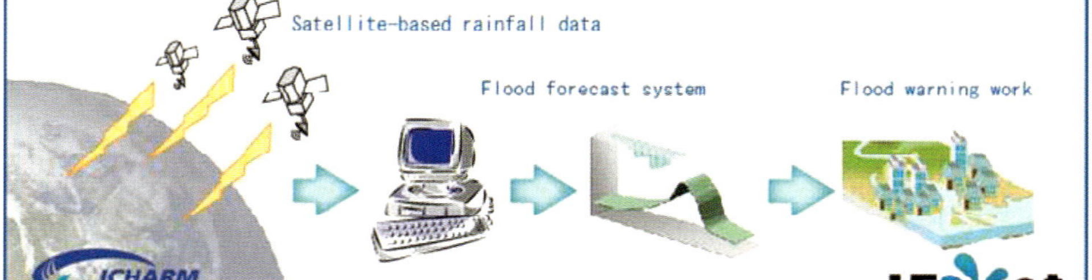

International Centre for Water Hazard and Risk Management (ICHARM)
International Flood Network (IF-Net)

Conclusion of INWEPF Workshop "Ecosystem Services of Paddy Irrigated Basin"
Dec 3, 2007, 14:00 – 17:00, New-life Plaza, Beppu

INWEPF

INWEPF WORKSHOP
"ECOSYSTEM SERVICES OF PADDY-IRRIGATED BASIN"

- **Background and Objectives**

International Network for Water and Ecosystem in Paddy Fields (INWEPF) is a network of paddy farming countries and international organizations to provide a forum to exchange knowledge and experiences of agricultural water management in paddy fields with due consideration for environmental aspects. Since its establishment in 2004, INWEPF has achieved several activities such as workshops, symposia, virtual meetings etc. At present, 17 countries mostly in Asia-Monsoon region participate in INWEPF.

"Water for Environment" is the important target of Priority Theme C of the 1st Asia-Pacific Water Summit, "Water for Development and Ecosystem." Water for Environment increases its importance in water management related to tackling recent climate change, conserving biodiversity and other issues on the basis of the concept of "Environmental Flows (E-flows)". Environmental services of paddy fields are also important issue for INWEPF, which addresses exchange of knowledge and experience about multiple features of paddy fields.

The workshop was organized to learn and discuss the experiences of application to the paddy irrigated basin with INWEPF member countries, related international organizations and participants.

- **Organizers** INWEPF Japanese Committee
- **Date & Venue** 14:00 – 17:30, December 3, 2007, New-life Plaza, Beppu City

- **Summary Report of Workshop**

The workshop started with 2 Keynote Speeches, "The Environmental Flows approach: Negotiating water management" by IUCN expert and "Environmental Flows: Moving from Theory to Application" by IWMI expert informing concept and actual application of E-flows.

The workshop had 2 speeches, "Hydrologic Environment in the Mekong River Basin and Roles of Paddy Fields" by NIRE Japan and "Channel Networking of Waters and Ecosystems in Paddy Irrigated Basins" by MAFF Japan to report environmental status and approach in Mekong River Basin and Japan. "Dialogue on Rice Paddies and Environmental Flows among INWEPF countries", a virtual meeting held prior to the workshop, was also reported by chairperson of the meeting.

Finally, "Message from INWEPF to the 1st Asia-Pacific Water Forum", which asserts importance of paddy farming and paddy irrigation water in order to address several issues in Asia, food security, poverty alleviation, conserving biodiversity etc.(See Annex).

The workshop resulted in exchange the recent information and opinions on environmental aspect related to surrounding paddy fields and irrigation. Participants fruitfully reconfirmed importance of paddy agriculture in tackling water security.

Annex: Conclusion of INWEPF Workshop "Ecosystem Services of Paddy Irrigated Basin"
Dec 3, 2007, 14:00 – 17:00, New-life Plaza, Beppu

Message from the INWEPF to the First Asia-Pacific Water Summit

INWEPF

Rice paddy water is essential for broad-based agricultural and rural development under rapidly increasing/changing demand for food in the world, increased competition for water from all sectors and accelerating global climate change. Meeting the Millennium Development Goals in the region requires improving food security, alleviating poverty, maintaining biodiversity, conserving the environment and promoting appropriate water resource management.

In the Asian monsoon region, where more than half of the world's population lives, rice is the most important staple crop. Rice paddy water is not only vital for food production, but also provides a broad range of services related to society, culture and the ecosystem, making an essential contribution to water security, which is the main goal of the 1st Asia Pacific Water Summit.

The International Network for Water and Ecosystem in Paddy Fields (INWEPF), a network of paddy farming countries and international organizations would like to convey the following important message to the Summit

(Efficient and Sustainable Water Use for Food Security and Poverty Alleviation)

The improvement of irrigation facilities and water management is important in order to improve the productivity and efficiency of rice paddy agriculture. It is also necessary to take actions to address climate change such as drought and flood that has an enormous effect on water resource management and threatens the capacity of ecosystems.

Governments should provide necessary assistance (e.g. policy, legal, financial and educational services and capacity building) for adequate investment, modernization and efficient management of rice water systems, in order to achieve food security, to alleviate poverty and to conserve ecosystems. Such programs should be supported to complement farmers' initiatives and should be geared towards improving the performance of irrigation management taking account of regional characteristics.

Modernizing irrigation management to enhance the welfare of farming communities, multiple roles and environmental sustainability of irrigation will require substantial improvements in system governance, flexibility, and management capacity.

(Multifunctionality and Maintaining Ecosystems)

Rice paddy water supports multiple functions that provide for fish cultivation, flood control, ground water recharge and culture heritage, as well as promote the conservation and maintenance of ecosystems. Paddy-systems serve as artificial wetlands which have good potential for preserving and maintaining both human cultures and natural ecosystems in a sustainable manner. Furthermore, rice paddy systems provide important ecosystems service at the river basin.

The multiple uses, roles, values, services and other aspects of agricultural water in paddy farming regions should be recognized, evaluated and incorporated into objectives, plans and strategies of water resources development and management.

(Promotion of Participatory Irrigation Management)

Paddy water system depends on agricultural water infrastructure and integrated water management systems, some of which have been developed through long-term traditional wisdom and experiences of local communities and their participation since the start of human history.

On introducing participatory irrigation management, the traditional wisdom and experiences of local communities and the value of their participation in water planning and management should be fully elicited. In view of accelerating irrigation sector reforms, we call for a new generation of participatory management strategies, focusing on demand-driven and service-oriented approaches and professionalization of management at all levels. In addition, the institution which supports farmers' activities and the functional framework which actually promotes the activities are indispensable.

Kyushu Regional Development Bureau, Ministry of Land, Infrastructure and Transport, Japan 3 December 2007

Waterborne Transportation and Regional Development

Executive Summary

Waterborne transportation (WT) is a mode of transport with less CO_2 emission compared to road transport. It was concluded that the further utilization of WT should be sought and implemented for promotion of regional development and cooperation, and effective countermeasures to mitigate and adapt climate change. On top of that, WT can be fully utilized for processing and recycling of waste materials by transporting those without affecting urban and rural traffic environment.

1. Background

Unit emissions of carbon dioxide by transport mode
(Unit: Ratio of CO2 emitted to transport 1 ton of cargo 1km)

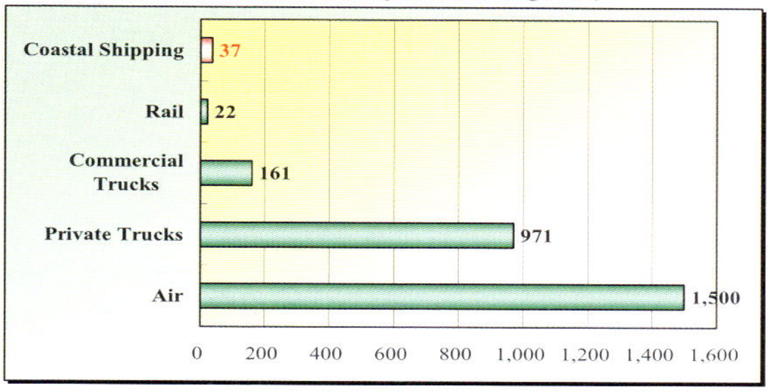

Source: Based on the Report on the Japan Greenhouse Effect Gas Inventory (October 2004) and Overall Energy Statistics (FY2003)

Waterborne transportation (WT, defined as ship transport through inland and coastal waterways) is a mode with low cost, energy efficient, and environment-friendly. At the Water and Regional Development Sessions of the 3rd and 4th World Water Forum held in Japan and Mexico, it was recommended that WT be further promoted and utilized to realize regional development. In this symposium, participants from Cambodia, Vietnam, Korea, China and Japan focused on advantages of WT in each region, and discussed on the measures to tackle challenges in further promoting WT.

2. Cases of waterborne transportation

(1) Mekong river basin

WT is utilized for both passenger and cargo transportation and supports the daily lives of people in local areas. Appropriate investment and management of waterways, such as navigational aids, GIS and River Information System, are necessary for the development of WT and sustainable economic and social development of the region.

(2) Northeast Asia

WT on the Seto Inland Sea and the Chang Jiang River are fully utilized for the trade between China, Korea and Japan. WT on the Chang Jiang River is important to stimulate the economy of the hinterland with less utilization of energy and traffic congestion. The Chinese government has a WT promotion policy which includes the adoption of the waterway regulation, dredging and vessel standardization.

(3) Japan

WT is utilized to transport waste materials and recycled materials between industrialized areas along the inland sea channel. This appears to be a new concept of utilizing WT. It is also utilized for tourism and for transport of emergency goods in case of disasters.

3. Conclusion

The participants of this symposium recognize as follows.

(1) Thanks to the blessing function of water, WT have been promoting regional development with relatively less energy consumption and with environmentally friendly manner. Further and safer operations of WT should be sought out.

(2) In developing WT systems, specific safeguards are required, such as standardization of vessels, maintenance dredging and provision of navigation aids, in order to utilize existing infrastructure.

(3) WT has comparative advantages to transport waste materials for processing and recycling, tourists and emergency materials during disaster.

(4) Experiences of each regional WT should be shared to each other to improve WT operations in each region. Further collaboration in Asia and Pacific Region should be sought.

WEPA
Water Environment Partnership in Asia

Pursuing a Sound Water Environment

- Summary of the 2nd International WEPA Forum - (Water Environment Partnership in Asia)

Water environmental governance needs to be improved for sound water environment

At the second International Forum of the Water Environment Partnership in Asia (WEPA), serious water pollution in Asia has been reconfirmed. A sound water environment is one of the essential elements of sustainable development in the region and there is a need for an improvement of water environmental governance.

The improvement of water environmental governance needs the accumulation and sharing knowledge and lessons, capacity development, and the strengthening of the network of relevant stakeholders working for water environmental conservation

Stakeholders from East and Southeast Asia representing government, academia, the private sector and NGOs gathered at the Forum. Through presentations and discussion, they shared experiences and knowledge with each other. Such sharing of knowledge and network of relevant stakeholders can be a driving force to promote good water environmental governance. WEPA will provide opportunities to share information and strengthen networks.

The 2nd International WEPA Forum

◆**Date:** 3 - 4 December 2007
◆**Venue:** Oita New Life Plaza, Beppu, Oita, Japan
◆**No. of Presentations:** 42 from East and Southeast Asian countries
◆**Organizer:** Ministry of the Environment, Japan
◆**Supporting Organizations:** Oita Prefecture, Beppu City, Ritsumeikan Asia Pacific University

Identified issues on water environment in East and Southeast Asia

- Water pollution is serious in rivers, lakes and groundwater caused by untreated or inadequately treated wastewater and poor sanitation facilities
- Importance of strengthening integrated water environment management, especially in transboundary water courses
- Importance of water policy development to cope with potential impacts of climate change
- Significance of introduction of policies and technologies suitable to the local situation and facilitation of public awareness

Identified actions to promote water quality improvement

- Strengthening of network among countries and also among water experts, and promotion of dialogue and capac ty building through such networks
- Enhancement of knowledge and information sharing through WEPA database
- Promotion of actions by each country for achievement of the Millennium Development Goals and subsequent information shar ng
- Installment of domestic wastewater treatment facility and sanitary system that is suitable for local conditions
- Promotion of water environmental conservation in view of integrated water resource management

The 2nd International WEPA Forum was held as an open event of the 1st Asia-Pacific Water Summit.

WEPA
Water Environment Partnership in Asia

Water Environment Partnership in Asia (WEPA)

Sound water environment is needed to ensure safe water access and sound environment. However, many Asian countries are faced with serious water pollution problems, which become a barrier to sustainable development in those countries.

Recognizing that effective water environmental governance is essential to solve such problems, the Ministry of the Environment of Japan launched the Water Environment Partnership in Asia (WEPA) in 2004.

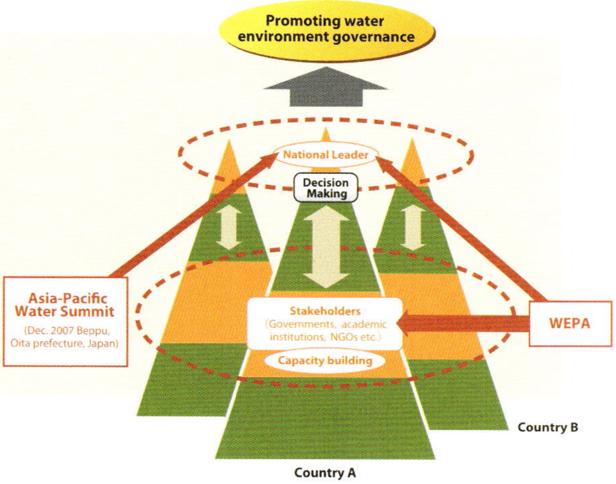

To promote effective environmental governance, WEPA activities are implemented under the partnership of 11 Asian countries, aiming to:

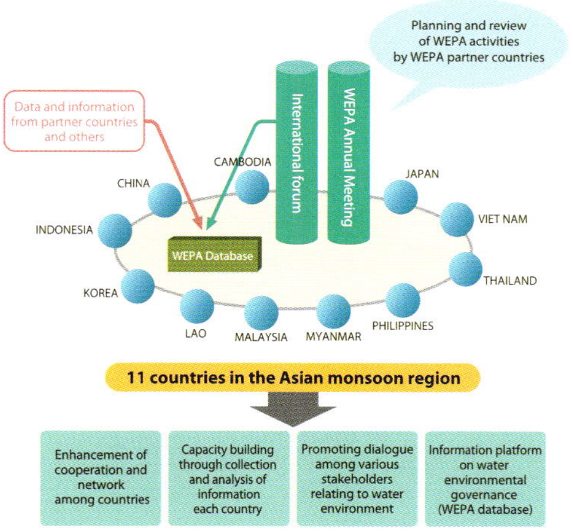

- enhance information exchange and cooperation on water environmental management among WEPA partner countries;
- promote dialogue and explore networks among governmental officials, researchers, NGOs and CBOs, and other relevant stakeholders;
- develop capacity of relevant stakeholders working for water environmental conservation through data collection and analysis, and
- provide a platform of information and knowledge on water environmental governance in the region as a form of WEPA database.

WEPA
Water Environment Partnership in Asia

WEPA Secretariat
c/o Freshwater Project, Institute for Global Environmental Strategies
2108-11 Kamiyamaguchi, Hayama, Kanagawa, 240-0115 Japan
Tel: +81-46-855-3880 Fax: +81-46-855-3809 E-mail: contact@wepa-db.net http://www.wepa-db.net/

PRINTED WITH SOY INK
R100

イベント名：「若者と有識者との対話セッション」
イベント主催者：若者と有識者との対話セッション実行委員会
開催日時：12 月4 日（火） 連絡先：hirokisato_japan@yahoo.co.jp

Agenda and proposal from Youths to Experts at the "Dialogue between Youths and Experts", a open-event at "Asia-Pacific Water Summit".

We the youth hereby proclaim to establish the Youth Disaster Reconstruction Volunteer Network with the rising awareness of disaster threats and in aim for the earliest possible reconstruction of the afflicted areas.

I. What We Decide (from agenda of the "Asia-Pacific Youths water summit")

1. We will exert our youth power to bring back the smiles on the faces of the victims and will actualize prompt support to revitalize the afflicted areas.
2. We will precisely identify the emergency information of the afflicted areas and will immediately spread information extensively.
3. We will be fully aware of the possibility of secondary disasters and will voluntarily hurry over to the afflicted areas.
4. We will also create the "Youth action map" that specifically identifies the relationship of various activities each youth is involved in the respective regions.
5. We will send out information pertaining to methods of disaster support and will provide more opportunities for information exchange.
6. We will communicate our know-how and thoughts to junior high and high school students. As a guest-teacher of disaster preventions at schools, or as a staff of a event for students, we would like to let them take over our activities.
7. For more quick cooperation with one another in emergency situations, we will provide opportunities for exchange on a regular basis where college students from various regions can get to know each other and will work hard to enhance the sharing of information and networking.

II. What we propose and what we request

A) For civic groups in basins

1. We request the opportunities that we can hold enjoyable events for information exchange.
2. We request civic groups in afflicted areas to supply local information.
3. We request civic groups to tell capable facilities when disasters are occurred.
4. We request skilled peoples to make training programs and hold workshops.
5. As a guest teacher and staff of the events for students, we request the opportunities that we can tell high school students and junior high school students about disaster preventions and water environmental activities.

B) For administrations, governments and enterprises

1. For youths' prompt support, we request cooperating groups and local groups to supply information about afflicted areas.
2. As a guest teacher and staff of the events for students, we request the opportunities that we can tell high school students and junior high school students about disaster preventions and water environmental activities.
3. We request administration, government and enterprise people in afflicted areas to supply materials which are needed for our activities.
4. We request skilled peoples to make training programs and hold workshops.
5. We request active utilization of "Youths action map". Moreover, we request experts' cooperation.
6. We request the "Water-related students' exchange program". Specifically, we request to facilitate youths' participation and make solid program, as well as financial and systematic cooperation, such as establishing the relationships among host countries.
7. To know secondary disasters, we request the supplement of the information about disasters.
8. We request to supply that we can hold activity exchange events in Japan in rotation. We hope that this youths' summit in Beppu is realized as a first youths' Asia-Pacific summit, and we wish that we are going to hold the "Asia-Pacific Youths' summit" in Japan and in the World again.

We the Youths promise these actions above.
To realize these actions, we hope that we can receive positive proposals and opinions from experts.